International Perspectives in Geography

AJG Library

Volume 23

Aim and Scope

The AJG Library is published by Springer under the auspices of the Association of Japanese Geographers. This is a scholarly series of international standing. Given the multidisciplinary nature of geography, the objective of the series is to provide an invaluable source of information not only for geographers, but also for students, researchers, teachers, administrators, and professionals outside the discipline. Strong emphasis is placed on the theoretical and empirical understanding of the changing relationships between nature and human activities. The overall aim of the series is to provide readers throughout the world with stimulating and up-to-date scientific outcomes mainly by Japanese and other Asian geographers. Thus, an "Asian" flavor different from the Western way of thinking may be reflected in this series. The AJG Library will be available both in print and online via SpringerLink.

About the AJG

The Association of Japanese Geographers (AJG), founded in 1925, is one of the largest and leading organizations on geographical research in Asia and the Pacific Rim today, with around 3000 members. AJG is devoted to promoting research on various aspects of human and physical geography and contributing to academic development through exchanges of information and knowledge with relevant internal and external academic communities. Members are tackling contemporary issues such as global warming, air/water pollution, natural disasters, rapid urbanization, irregular land-use changes, and regional disparities through comprehensive investigation into the earth and its people. In addition, to make the next generation aware of these academic achievements, the members are engaged in teaching and outreach activities of spreading geographical awareness. With the recent developments and much improved international linkages, AJG launches the publication of the AJG Library series in 2012.

Takashi Yamazaki · Godfrey Baldacchino
Editors

Relational Geographies of Islands in the Indo-Pacific

Environment, Society, and Politics

Springer

Editors
Takashi Yamazaki
Graduate School of Literature and Human Sciences
Osaka Metropolitan University
Osaka, Japan

Godfrey Baldacchino
Department of Sociology
University of Malta
Msida, Malta

ISSN 2197-7798 ISSN 2197-7801 (electronic)
International Perspectives in Geography
ISBN 978-981-96-7154-0 ISBN 978-981-96-7155-7 (eBook)
https://doi.org/10.1007/978-981-96-7155-7

This work was supported by the Japan Society for the Promotion of Science (18KK0029).

This Springer imprint is published by the registered company Springer Nature Singapore Pte Ltd.
The registered company address is: 152 Beach Road, #21-01/04 Gateway East, Singapore 189721, Singapore

Preface

This book is a collection of papers by authors who participated in the International Geographical Union (IGU) Thematic Conference on "Islands in Relations: Conflicts, Sustainability, and Peace," held on April 4–7, 2023, at Osaka Metropolitan University, Osaka, Japan (see https://polgeog.jp/igu-tc2023/). The Conference attracted 121 participants from 28 countries, and comprised 6 invited keynote speeches and 86 general presentations in 27 sessions. One of the authors of this Preface (Yamazaki) was the chief organizer, and the other (Baldacchino) was one of the keynote speakers. In this Preface, we introduce the objectives of the Conference, briefly review the contemporary challenges that islands face, and provide an overview of the book chapters.

The Conference took note of the various definitions and provisions that apply to islands, including the United Nations Convention on the Law of the Sea (UNCLOS), and decided to focus on and address three component themes. First is the theme of power relations. Islands have various forms and social, economic, and political relations with other islands and continents. While some islands may constitute a state, others may be dominated by neighboring islands or continental states. Islands are situated in a complex web of sovereign powers. This situation may be called "mainland/island relations" (Mountz 2015), exemplified by colonization, territorialization, militarization, or marginalization. Conflicts can emerge along these lines between states over islands (e.g., territorial disputes or wars) or between the mainland and an island (e.g., forcible annexation or separatism). The Conference investigated why and how islands were/are involved in what kind of conflicts or uneven power relations.

Second is the theme of islandness (the condition of being an island). Islands are geographically characterized by oceanity, remoteness, and smallness. While these aspects may lead to a socio-economic vulnerability, they can also open islands toward fisheries, exclusive tourism, inter-island networks, and maritime trade. Islands' unique climate, geomorphology, or ecology can also create distinctive cultures, local products, endemic forms of life, and tourist attractions. In this sense, sustainability is crucial for the islands' development. However, "mainland/island relations" may distort such development towards dependence on the mainland's investment, consumption, or public finance, increasing the mainland's effective control and purview over an island. The Conference examined how islands' sustainable development can contribute to their political-economic autonomy.

Third, islands are battlegrounds. Islands are often portrayed or assumed to be tranquil spaces; but many serve as battlefields for and between imperial powers. Some remain under heavy military presence. Such islands have no peace but putatively provide peace for their respective mainlands. Is geography part of this quandary, and how could one deploy geography for island peace? Answers may be twofold: epistemological and practical. Epistemologically, we may be caught in "the territorial trap" (Agnew 1994) even when we see islands. To assure their sustainability, islands must connect to and with other islands and continents through and across seas, and these human-geographical elements and trans-aquatic entanglements and networks constitute a "liminal" (in-between) space; a milieu for transnational connectivity (Matsuda 2019). A territorially trapped epistemology needs to be deconstructed. Practically, state borders marginalize border islands while trans-border interactions such as border tourism, sister-city exchange, or bilateral developmental initiatives for maritime borderlands can increase islands' sustainability while avoiding (and even reducing) tensions. Geography can contribute towards promoting such interactions and empowering island peoples.

While there are various regional frameworks for understanding international relations geographically, the Osaka Conference focused on islands by re-situating the world's many islands in their historical, geographical, cultural, economic, and political relationships with their surrounding regions (other islands, mainland, oceans, continents), understanding their past and present, and looking towards the future. As symbolized by its subtitle (conflict, sustainability, and peace), the Conference reflected on the history of the islands, which have often been at the mercy of hegemonic domination by major powers and their military adventures and confrontations. Participants actively discussed ways to achieve political and economic independence as well as environmental and cultural sustainability of the islands, seeking a durable and stable peace and security. The 3-day post-conference field trip to Okinawa illuminated these Conference topics empirically on the island as a US military outpost.

The frameworks mentioned above are directly reflected in the title of this book, *Relational Geographies of Islands in the Indo-Pacific: Environment, Society, and Politics.* Island studies scholars have presented various relational frameworks such as mainland/island relations, hybrid island epistemology (Baldacchino 2008), archipelagic thinking (Mountz 2015), relationality (Pugh 2016), Anthropocene thinking, and abyssal geography (Chandler & Pugh 2022). Islands are not isolated or self-contained territorial units but are situated in ever-changing inter-relationships between humans and non-human environments, intra- and extra-island dynamics, and lands and seas. "Islands in Relations" properly represents the relational reality of island geographies.

In the next section, based on the papers and discussions shared at the Conference, we will briefly highlight the current challenges islands face and the options or countermeasures they can choose.

Serious concerns about the state of health of Planet Earth, as well as lingering postcolonial geopolitics, are forcing a re-evaluation of the "nature-culture" nexus across all disciplines (Bhagat-Kennedy 2018; Roth 2021). Thus, a sound theoretical and empirical understanding of the changing relationships between nature and human activities is a hallmark of contemporary geographical pursuits. The manner in which space is reconfigured in the current epoch owes much to overarching forces at work, including nationalism and identity politics, globalization, neo-colonialism and re/de-militarization. Singly and collectively, humans resist and respond to the ensuing opportunities and threats, where and how they can: migration, domestic and international, are significant responsive measures. Of particular interest to the Osaka Conference is how islands shape up to be archetypal sites of/for specific relational geographies: the myriad ways in which people construct and navigate lived spaces when their communities are entangled in the dimensions of islandness: smallness, boundedness, embraced by water, separated from the mainland (Baldacchino 2018; Kelman 2020).

Islands have long attracted the attention of geographers (King 1999). In the current era, islands have become alternating geographies of hope (Turner 2008) and despair, providing examples, from real life and in real time, of the consequences of either misguided or farsighted development. From the eco-island initiative of Ishigaki, Okinawa Prefecture, Japan (Kakazu 2018); to the toxic "garbage island" of Thilafusi, Maldives (Munshi et al. 2022).

We argue that the stretching of material and social relations across and through spaces on islands is quite particular. Indeed, small islands are characterized by a combination of sudden, rapid and total consequences of booms and busts, peaks, and troughs. Islanders must engage with the relational geographies emergent from these three conditions. When experiencing such high levels of elasticity, no wonder then that islands—sometimes the *same* islands, at different epochs—can face two extreme scenarios: isolation and abandon on one hand; incorporation and co-optation on the other.

A simple contrast has been presented in the literature between two historical island roles, and which largely match these two extreme predicaments (Connell & King 1999). First, islands as *marginal* spaces, which tend eventually to become areas of outmigration and depopulation, leading to the MIRAB syndrome, where the locals somehow eke a meagre survival from (domestic and/or international) aid disbursements and remittances (the latter fueled by outmigration, which can be actively encouraged), with some "make work" public sector employment at home (Bertram & Watters 1985). Second, islands as *nodal* places. Such islands tend to attract and exchange populations, leading to the creation of cosmopolitan, hybrid and stratified societies, often with a more "open" mentality towards the outside world. Here are the *entrepôt* islands, which benefit from strategic proximity to trade routes, functioning as semi-detached entry points for contiguous mainlands. The entrepôt island trades on talent, commerce, and investment rather on large-scale migration; a combination of emigration and immigration, especially of wealthy merchants, entrepreneurs and tax exiles, is to be expected (King 1999, 2009). Port cities, endowed by natural harbors, provide the geography to catalyze islands as entrepôts (Darwin 2020), as also argued by Chapter "Island Ports: Insights into the 'Betweenness' of Island Spaces" in this volume. Of course, their natural endowments also made such islands attractive assets to colonial powers (Warrington & Milne 2018). No wonder many small islands have been among the first territories to be colonized, and are amongst the last to remain so (Connell & Aldrich 2020; Feyrer & Sacerdote 2009).

The signature of island isolation and abandonment involves (especially youth) outmigration, lack of investment, an aging island community, limited economic and entertainment services, and a loss of hope in a viable and secure future. Industrialization privileged cities and economies of scale; small islands could not compete with this trend and succumbed to "vicious cycles of decline" (Baldacchino & Starc 2021). And, in the twenty-first century, one major agent behind such a dynamic is the palpable consequence of climate change. A combination of sea level rise, ocean acidification, coral bleaching, unsustainable fishing practices, saltwater intrusion into water tables, more violent storms and higher tides is traceable to an exploitative and polluting model of development which has been a paradigm of modernity, and the posterchild of the Age of the Anthropocene (Mycoo et al. 2022; Petzold & Magnan 2019).

Where small islands are concerned, including small island developing states (SIDS), the prognostications are stark. We are already witnessing the suddenness of impact, the intensity of effect, and the speed of penetration of climate change and its consequences. Whole island communities have already been displaced; whole islands have already been submerged (Connell 2018a). Mitigation strategies, while laudable, have been lethargically implemented; they have also met stiff resistance from vested economic and political interests. Pious hopes, redolent of the optimism of the Paris 2015 Summit, to thwart the rise in the mean temperature of the planet to not more than 1.5@ when compared to pre-industrial levels, are now accepted as having been overly optimistic: the aspiration is, frankly, obsolete (Cointe & Guillemot 2023).

For small islands, unlike other territories, there is simply no strategic depth; no "hinterland to retreat to" (Hintjens 1991: 38). The policy options really boil down to two: aggressive adaptation for those who wish to remain and exercise the "right to stay"; and wholesale emigration and resettlement to suitable locations, in the immediate region or far away, for those who have given up on staying. This is why island communities "left behind" are facing an existential challenge. They are desperate to be proactive about their predicament while holding the polluting countries to task for causing the current climate mess in the first place. Indeed, it would be fair to say that SIDS—via the Alliance of Small and Island States (AOSIS)—have become visible on the international stage precisely by making a strong and sustained case in favor of curbing the excesses and negative externalities of the carbon age and championing sustainable development instead (Baldacchino & Antat 2023; Klöck et al. 2024).

In the chapters that grace this collection, four broad initiatives in contemporary island relationality are identified and presented. Each in their own way, they represent a very particular unfolding of political (structure versus agency) and environmental (land versus sea) dynamics with islands, or island states, and their peoples and governments, at their center. These are (1) an aggressive, proactive adaptation towards the consequences of climate change; (2) the option and decision to emigrate; (3) the provision of detention, military or security services; and (4) serving as advance fortresses and projections of hegemonic power. We review each of these below.

Aggressive adaptation: Initiatives in favor of aggressive adaptation at home include shoring beaches and planting mangroves to reduce the rate of coastal erosion, land reclamation, as well as constructing wholesale artificial islands to accommodate sections of the local population that are threatened by sea level rise, or who are simply migrating domestically to capital cities (Abdulla 2020; Chee et al. 2017). In this book, Chapter "Construction of the Artificial Island of Hulhumalé as a Climate Change Adaptation Measure in the Maldives" (about Hulhumalé, off Malé, Kaafu Atoll, The Maldives) and Chapter "Baseco Compound: The River Island of Contestation, Despair, and Hope in the Philippines" (about Baseco, in Metro Manila, The Philippines) speak to these island realities in the Indo-Pacific.

Outmigration: Migration has always been, and will remain, a vibrant feature of small island life (King et al. 1995). Island diasporas are everywhere; many times, there are more populous than the island communities from which they originally sprung (Spickard et al. 2002). They are responses to "push factors" such as chronic unemployment, underdevelopment, tense community surveillance, political discrimination, and the collapse in habitability occasioned by the impacts of climate change at home; and to such "pull factors" as quality education, more decent work options and an overall better quality of life that one aspires to elsewhere, and without existential threats brought about by failed harvests, hurricane damage or fickle economic activity (King 2009; Connell 2018b; Alexander & Henderson 2024).

Diasporas can and do take root and flourish in far-flung locations, possibly building island-island relations: consider, for example, the Sri Lankans in Sicily, Italy, as illustrated and examined by Chapter "From Island to Island: Sri Lankans' Settlement Model(s) in the Three Main Cities of Sicily, Southern Italy". Kiribati, one of four SIDS imminently and totally compromised by sea level rise, has actively pursued resettlement efforts by its citizens in other neighboring Pacific countries, including Fiji and New Zealand (Hermann & Kempf 2019): check out Chapter "Countermeasures to Climate Insecurity of Asia-Pacific Island Countries: Out-Migration and Adaptation Options" for more details. The same chapter included insights on the migration of citizens from the Marshall Islands—another of the compromised SIDS quartet—to the USA. The Marshall Islands, like Palau and the Federated States of Micronesia, while full members of the United Nations, are nevertheless unique "hybrid jurisdictions" (Levine & Roberts 2005) which have delegated some of their sovereign powers to the US in exchange for benefits that include freedom of movement to the US mainland. Such a wild card is a precious resource to islanders in dangerous and unpredictable times. Hence, the rationale for an enduring "US hegemony" in the Pacific (Pöllath 2018).

Border islands: And here we shift to a consideration of islands that are far from forgotten or compromised by outmigration. Instead, these islands continue to deliver services into the twenty-first century and are often the sites of particular influxes: of soldiers and their materiel; of tourists; of criminals, terrorists, and those who fall foul of the justice or immigration systems, members of an "enforcement archipelago" (Mountz 2011). As border islands, they provide facilities for the enactment of particular regional geographies. With an eye to the chapters of this collection, these services range from detention facilities and securitization to militarization, and offer a completely different expression of "development."

A case in point is the border island of Lieyu, a small island in Taiwan's Kinmen archipelago, 288 km from mainland Taiwan, but less than 16 km from the city of Xiamen, People's Republic of China. Long-term military restrictions and inconvenient connectivity within the de facto state of Taiwan/Republic of China have stunted Lieyu's development and has resulted instead in a unique natural and battlefield landscape (Lee et al. 2017). Here is a distinct island metageography (Baldacchino & Tsai 2014), with public opinion largely prepared to accept connective infrastructure—including a bridge to Kinmen—to improve their overall quality of life. The islet, like neighboring Kinmen, is so imbricated in its military history that it specializes in battlefield tourism (Zhang 2010).

The Indo-Pacific is the world's richest and largest populated archipelagic space. Island chains defy national frontiers and island states can illegalize migrant flows, transforming border islands into detention centers for such migrants (see Chapter "'Illegal' Korean Immigrants in Island(ed) Detention Camps and Enduring Colonial Sovereignty: The Case of Sasebo, Japan, Under U.S. Occupation"). Lingering colonial ties mean that various bits and pieces of these chains are administered by non-regional powers: the US primarily but also France and the UK (the latter mainly through notionally administering the Chagos Islands, claimed by Mauritius). Both China and India are looking to expand influence over the region, and cultivating good relations with small island states is an easy manifestation of soft power diplomacy (Beck 2020). For countries that are exclusively islands—and there are many—this situation carries the potential to concoct particular border tensions. Should neighbors not be friendly, then outlying islands on the geophysical and psychological edge of the nation become re-proposed and represented by the state as essential components of the motherland, a first and critical line of defense (Glaser 2022; Iwashita 2012). These in-between locales also function as intercultural zones, spaces where border tourism can flourish: surely a better prospect for island economic development than serving as a dumping site for nuclear waste. Chapter "Developing Border Tourism for the Sustainability of Border Island: The Case of Tsushima Island in Japan" explores this issue with regards to the border (Japan-Korea) island of Tsushima.

Fortress islands: The United States so far maintains an impressive military presence in the Western Pacific, with a string of allied countries and its own unincorporated territories. The aftermath of the Second World War also saw the US occupy Japan and assume its administration; it continued to govern Japan's southernmost prefecture of Okinawa even after its administration over the rest of Japan was concluded in 1952. The Okinawa archipelago hosts 70% of US military bases in Japan and is considered to be the lynchpin of its advanced defense system, with China as the presumed main rival (although Russia and North Korea cannot be excluded) (Oshiro 2024). The residents of Okinawa, meanwhile, must deal with the consequences of militarization. Seen through time, these involve contemporary moments of demilitarization—as certain military bases get decommissioned, often after sustained public outcry—and remilitarization, as new ground is broken to accommodate new bases elsewhere on the island. Rich details on this are offered by Chapter "Dual Militarization of the Pacific Island: Continuing Colonial Transformation of Okinawa, Japan".

Living alongside the military installations of a foreign power, with the full connivance of one's national government, has its consequences. Sure, there are benefits: a state-of-the-art infrastructure, and well-paying jobs, to the extent that some locals cannot even imagine a non-fortress economy. Then there are costs. These include pollution and contamination, issues often shrouded in secrecy and which illustrate the impotence of local (municipal) government (for more, see Chapter "Militarized Territoriality: Water Contamination and Military Operations in the Pacific").

The suddenness of impact, the intensity of effect, and the speed of penetration of extra-territorial military interests has created an "archipelago of island fortresses" around the world: notably, Guam, Hawai'i and Okinawa in the Pacific; and Diego Garcia in the Indian Ocean. The presumed immediate threat here is *not* climate change.

In light of these complex contemporary challenges and the options that islands have to deal with, the book is organized into the following nine chapters. The first chapter "Island Ports: Insights into the 'Betweenness' of Island Spaces," by Godfrey Baldacchino, presents the global historiographies of island ports and reveals that such ports have developed as an interface between the inside and outside of the island, rendering its space liminal in terms of transportation, economy, and culture. However, this is not a one-directional process for island openness; there have been moves towards the opposite. This "betweenness" of islands makes their geographies relational; thus, this chapter acts as an introduction to the whole book.

The second chapter "Construction of the Artificial Island of Hulhumalé as a Climate Change Adaptation Measure in the Maldives," by Akiko Sakamoto et al., and The third chapter "Baseco Compound: The River Island of Contestation, Despair, and Hope in the Philippines," by Joefe B. Santarita and Godfrey Baldacchino, are examples of *aggressive adaptation* (i.e., the construction of artificial islands) towards drastically changing environments, as mentioned above. The second chapter "Construction of the Artificial Island of Hulhumalé as a Climate Change Adaptation Measure in the Maldives" looks at the Maldives, a naturally low-lying archipelago in the Indian Ocean, where the artificial island of Hulhumalé has been constructed as a climate change adaptation measure. The chapter explores the socio-economic and environmental aspects of Hulhumalé's development by analyzing residents' perspectives on living conditions, migration patterns, and other social costs caused by such a massive project. The third chapter "Baseco Compound: The River Island of Contestation, Despair, and Hope in the Philippines" investigates the case of Baseco Compound, an artificial river island in Manila, the capital of The Philippines, and illustrates the poignant history, geography, and challenging socio-economic conditions of its urban islanders.

The option of *outmigration* from islands in the Indo-Pacific is discussed in the fourth chapter "From Island to Island: Sri Lankans' Settlement Model(s) in the Three Main Cities of Sicily, Southern Italy," by Francesca Bitonti et al., and the fifth chapter "Countermeasures to Climate Insecurity of Asia-Pacific Island Countries: Out-Migration and Adaptation Options" by Juha I. Uitto et al. The fourth chapter "From Island to Island: Sri Lankans' Settlement Model(s) in the Three Main Cities of Sicily, Southern Italy" conducts qualitative analyses to compare Sri Lankans' urban settlement patterns in Sicily and to detect possible spatial polarization and socio-economic inequalities of Sri Lankans at the intra-urban level. The fifth chapter "Countermeasures to Climate Insecurity of Asia-Pacific Island Countries: Out-Migration and Adaptation Options", on the other hand, explores possible means to sustain SIDS against climate-induced risks in the Asia-Pacific.

The sixth chapter "'Illegal' Korean Immigrants in Island(ed) Detention Camps and Enduring Colonial Sovereignty: The Case of Sasebo, Japan, Under U.S. Occupation," by Taku Fukumoto, and the seventh chapter "Developing Border Tourism for the Sustainability of Border Island: The Case of Tsushima Island in Japan," by Yasunori Hanamatsu, investigate *border islands* in Japan. The sixth chapter "'Illegal' Korean Immigrants in Island(ed) Detention Camps and Enduring Colonial Sovereignty: The Case of Sasebo, Japan, Under U.S. Occupation" focuses on the postwar internment camps for "illegal" Korean immigrants, isolated at Sasebo Port, to illuminate the persistence of colonial elements in state sovereignty, which emerges as border control, detention, and deportation even today. In contrast, the seventh chapter "Developing Border Tourism for the Sustainability of Border Island: The Case of Tsushima Island in Japan" presents the current situation of Tsushima to show how the island is located at a crossroads between sustainable development as a space for border tourism and its uncertain future as a site of nuclear waste disposal.

The last two chapters address Okinawa to demonstrate the dynamic of *fortress islands*. The eighth chapter "Dual Militarization of the Pacific Island: Continuing Colonial Transformation of Okinawa, Japan," by Takashi Yamazaki, characterizes the US military presence on the island as an imperialist project that perpetuates US hegemonic status at the expense of islanders' autonomy. It does so by analyzing the interlinkage between de-militarization and re-militarization processes in Okinawa. The ninth chapter "Militarized Territoriality: Water Contamination and Military Operations in the Pacific," by Tatsuki Kohatsu, uses episodes of water contamination by US military bases to argue that US geopolitical power is contingent on the way its military maintains its smooth operations on the ground across the Pacific while the lives of those on the impacted islands provide critical and insightful reflections on the US' workings.

Island geographies are relationally constructed to become an assemblage of environmental, social, and political dynamism in the Indo-Pacific, a vast space in which islands continue to be key sites for connecting continents and oceans. The arguments and cases that grace *Relational Geographies of Islands in the Indo-Pacific: Environment, Society, and Politics* provide a powerful picture of the significance and centrality of islands in the twenty-first century, far from an easily assumed benign peripherality and invisibility.

Osaka, Japan
Msida, Malta
November 2024

Takashi Yamazaki
Godfrey Baldacchino

Acknowledgments The editors are grateful to the Japan Society for the Promotion of Science (Project No. 18KK0029), the Tokyo Geographical Society, the International Geographical Union, the Association of Japanese Geographers, the Human Geographical Society of Japan, and the Osaka Convention and Tourism Bureau for financially supporting the 2023 IGU Thematic Conference in Osaka, which is the basis for this book project and its contents.

References

Abdulla AA (2020) Sea level rise, statehood and artificial islands for environmentally displaced persons under international law: future of the Maldives. Int J Soc Res Innov 4(1). https://doi.org/10.55712/ijsri.v4i1.21
Agnew J (1994) The territorial trap: the geographical assumptions of international relations theory. Rev Int Political Econ 1(1):53–80
Alexander R, Henderson H (eds) (2024) Higher education in small islands: Challenging the geographies of centrality and remoteness. London: Policy Press
Baldacchino G (2008) Studying islands: on whose terms? Some epistemological and methodological challenges to the pursuit of island studies. Island Studies Journal 3(1):37–56. https://doi.org/10.24043/001c.81189
Baldacchino G (2018) Preface. In G. Baldacchino (ed), Routledge international handbook of island studies: A world of islands (pp xix–xxxv). London: Routledge
Baldacchino G, Antat S (2023) Degrees of success: the alliance of small island states at the 2015 Paris climate change conference. In G. Baldacchino (ed), The success of small states in international relations: Mice that roar? (pp 124–135). London: Routledge
Baldacchino G, Starc N (2021) The virtues of insularity: pondering a new chapter in the historical geography of islands. Geography Compass, 15(12): e12596. https://doi.org/10.1111/gec3.12596
Baldacchino G, Tsai HM (2014) Contested enclave metageographies: the offshore islands of Taiwan. Political Geography 40(1):13–24. https://doi.org/10.1016/j.polgeo.2014.02.002
Beck C (2020) Geopolitics of the Pacific Islands. Security Challenges 16(1):11–16. https://www.jstor.org/stable/26908762. Accessed 24 September 2024.
Bertram IG, Watters RF (1985) The MIRAB economy in South Pacific microstates. Pacific Viewpoint 26(3):497–519. https://doi.org/10.1111/apv.263002
Bhagat-Kennedy M (2018) Nation after world: rethinking 'the end of postcolonial theory.' Interventions 20(3):335–344. https://doi.org/10.1080/1369801X.2018.1452625
Chandler D, Pugh J (2022) Interstitial and abyssal geographies. Political Geography 98. https://doi.org/10.1016/j.polgeo.2022.102672

Chee SY, Othman AG, Sim YK, Adam ANM, Firth LB (2017) Land reclamation and artificial islands: walking the tightrope between development and conservation. Global Ecology and Conservation 12(1):80–95. https://doi.org/10.1016/j.gecco.2017.08.005
Cointe B, Guillemot H (2023) A history of the 1.5 °C target. Wiley interdisciplinary Reviews: Climate Change 14(3):e824. https://doi.org/10.1002/wcc.824
Connell J (2018a) Nothing there atoll? Farewell to the Carteret Islands. In T. Crook & P. Rudiak-Gould (eds), Pacific climate cultures: Living climate change in Oceania (pp 73–87). Warsaw: De Gruyter Open Access
Connell J (2018b). Migration. In G. Baldacchino (ed), Routledge international handbook of island studiues: A world of islands (pp 261–278). London: Routledge
Connell J, Aldrich R (2020) The ends of empire. Singapore: Springer Singapore
Connell J, King R (1999) Island migration in a changing world. In R. King & J. Connell (eds), Small worlds, global lives: Islands and migration (pp 1–26). London: Pinter
Darwin J (2020) *Unlocking the world: Port cities and globalization in the age of steam, 1830–1930*. London: Penguin
Feyrer J, Sacerdote B (2009) Colonialism and modern income: islands as natural experiments. The Review of Economics and Statistics 91(2):245–262. https://doi.org/10.1162/rest.91.2.245
Glaser BS (2022) Conflict in the South China Sea. New York: Council on Foreign Relations. http://www.jstor.com/stable/resrep05645. Accessed 24 September 2024
Hermann E, Kempf W (2019) Adaptation and the question of migration: directions in dealing with climate change in Kiribati. In C. Klöck & M. Fink (Eds.), *Dealing with climate change on small islands: Towards effective and sustainable adaptation* (pp 293–312). Göttingen: Göttingen University Press. https://library.oapen.org/bitstream/handle/20.500.12657/23566/1/1006579.pdf#page=307. Accessed 24 September 2024
Hintjens HM (1991) France in the Caribbean. In P. Sutton (ed), Europe and the Caribbean (pp 37–70). New York: Macmillan
Iwashita A (2011) An invitation to Japan's borderlands: at the geopolitical edge of the Eurasian continent. Journal of Borderlands Studies 26(3):279–282. https://doi.org/10.1080/08865655.2011.686969
Kakazu H (2018) Island sustainability and inclusive development: the case of Okinawa (Ryukyu) Islands. J Mar I Cultures 7(2): 1–36. https://jmic.online/issues/v7n2/1/. Accessed 24 September 2024
Kelman I (2020) Islands of vulnerability and resilience: Manufactured stereotypes? Area 52(1):6–13. https://doi.org/10.1111/area.12457
King R (1999) Islands and migration. Biagini E Hoyle B (eds), Insularity and development: International perspectives on islands (pp 93–115). London: Pinter
King R (2009) Geography, islands and migration in an era of global mobility, Island Studies Journal, 4(1):53–84. https://doi.org/10.24043/isj.228
King R, Connell J, White PE (eds) (1995) Writing across worlds: Literature and migration. London: Psychology Press
Klöck C, Castro P, Chin-Yee S, Gurresø C, Desmasures C, Bouly E, de Pryck K (2024) Beyond AOSIS: small island states' presence and participation at COP27. Climate and Development, 1–12. https://doi.org/10.1080/17565529.2023.2298780
Lee SH, Huang WH, Grydehøj A (2017) Relational geography of a border island: local development and compensatory destruction on Lieyu, Taiwan. Island Studies Journal, 12(2):97–112. https://doi.org/10.24043/isj.33.
Levine S, Roberts NS (2005) The constitutional structures and electoral systems of Pacific island states, Commonwealth and Comparative Politics, 43(3):276–295. https://doi.org/10.1080/14662040500304866
Matsuda, H. 2019. Liminality of the Japanese Empire: Border crossings from Okinawa to colonial Taiwan. Honolulu: University of Hawai'i Press.
Mountz A (2011) The enforcement archipelago: Detention, haunting, and asylum on islands. Political Geography, 30(3):118–128. https://doi.org/10.1016/j.polgeo.2011.01.005

Mountz A (2015) Political geography II: Islands and archipelagos. Progress in Human Geography 39(5):636–46. https://doi.org/10.1177/0309132514560958
Munshi S, Banerjee S, Chakraborty I (2022) Thilafushi a toxic bomb: Learning from an island's solid waste management practice. *International Journal of Architecture and Infrastructure Planning* 8(2):1–12. https://doi.org/10.37628/jaip.v8i2.954
Mycoo M, Wairiu M, Campbell D, Duvat V, Golbuu Y, Maharaj S, Warrick O (2022) Chapter 15: Small islands. IPCC WGII Sixth Assessment Report. https://hal.science/hal-03704367/document Accessed 24 September 2024
Oshiro A (2024) From "footprint" to relationships: Impacts of US military base on Okinawa. Sociology Compass 18(1):e13099. https://doi.org/10.1111/soc4.13099
Petzold J, Magnan AK (2019) Climate change: thinking small islands beyond Small Island Developing States (SIDS). Climatic Change, 152(1):145–165. https://doi.org/10.1007/s10584-018-2363-3
Pöllath M (2018) Revisiting island decolonization: the pursuit of self-government in Pacific island polities under US hegemony. Island Studies Journal 13(1):235–250. https://doi.org/10.24043/isj.46
Pugh J (2016) The relational turn in island geographies: Bringing together island, sea and ship relations and the case of the Landship. Social and Cultural Geography 17(8):1040–1059. https://doi.org/10.1080/14649365.2016.1147064
Roth S (2021) The great reset: restratification for lives, livelihoods, and the planet. Technological Forecasting and Social Change 166:120636. https://doi.org/10.1016/j.techfore.2021.120636
Spickard P, Rondilla JL, Wright DH (eds) (2002) Pacific diaspora: Island peoples in the United States and across the Pacific. Honolulu: University of Hawai'i Press
Turner C (2008) The geography of hope: A tour of the world we need. Toronto: Vintage Canada
Warrington E, Milne D (2018) Island governance. In G. Baldacchino (ed) The Routledge international handbook of island studies: A world of islands (pp 173–201). London: Routledge
Zhang JJ (2010) Brand(ing) Kinmen: a tourism perspective. Int J Entrep Small Bus 9(4):407–433. https://doi.org/10.1504/IJESB.2010.032402

Contents

Island Ports: Insights into the 'Betweenness' of Island Spaces

Godfrey Baldacchino

Abstract The actual personality and character of islands as liminal spaces, at once local and global, at once national and cosmopolitan, may depend significantly on the dynamics of their port (and often capital) cities. By looking at what happens in the spaces of potential cultural exchange and brokerage in such locales—including working waterfronts, markets and commercial transactions, coastal tourism establishments and activities, religious rituals and celebrations—one gets a better sense of whether and how islands, in spite (or because) of their smallness, can be ripe for human conviviality, multilingualism, tolerance and a peace based on prosperity. In contrast, it is when island societies develop a narrow, exceptionalist and special definition of themselves—including pillarised notions of their own, internal island communities in terms of race, ethnicity and nationhood—and become convulsed by questions of identity, that they reduce their appeal as 'welcoming societies' and instead become the strategic bulwarks, flanks and edges of hard borders.

Keywords Islands · Port cities · Capital cities · Liminal spaces · Entrepôts

1 Introduction

> The port city ... represents a strange political condition.
>
> Human societies usually define themselves by their norms and rules [*nomos*]... Yet, in the port city, governed by the pendulum motion of the sea and antinomical rhythms of arriving and departing guests, the hold of *nomos* on human identity seems that much weaker... In a world constituted by the oscillating movement between ports of call, it is difficult to institute one common set of rules, techniques, coinage, language or customs. Contradictory norms find ready acceptance.
>
> (Murphy 2001, *passim*)

G. Baldacchino (✉)
Department of Sociology, University of Malta, Msida, Malta
e-mail: godfrey.baldacchino@um.edu.mt

T. Yamazaki and G. Baldacchino (eds.), *Relational Geographies of Islands in the Indo-Pacific*, International Perspectives in Geography 23,
https://doi.org/10.1007/978-981-96-7155-7_1

The author of this chapter is someone born and bred on the Mediterranean archipelago of Malta, the tenth smallest country in the world by land area. Hence, his obligation to ponder: why is Malta a sovereign state and a labour importing economy, with a high population density; while its neighbouring Italian islands of Pantelleria, Linosa and Lampedusa, with a similar climate, geology and topography, have very small populations, attract limited investment and are experiencing depopulation (Baldacchino 2015)? Part of the answer lies in Malta's Grand Harbour, a natural, deep water, all-weather natural asset that has made the Maltese Islands attractive to all Mediterranean powers but also as an obvious starting point to develop port infrastructure. None of the Italian island trio has such an advantage.

Darwin (2020) describes 'port cities' as 'gateway cities', springing up on the edge where two different zones meet. They are mainly maritime cities: the place to exchange the produce of different economies and cultures; and where goods are transferred from one mode of transport to another. They are the home of ancillary services: shipping agents, insurers, money-changers, banks, hauliers and draymen, dockers and porters, pimps and prostitutes, and the merchants who manage the transactions. Some port cities dealt in *staples*, and therefore collected and exported the produce of their own immediate neighbourhood, and which depended on the availability of a hinterland close by. Meanwhile, other port cities were *entrepôts*: lacking their own hinterland, their commercial relations were with similar ports located elsewhere, and with distant suppliers and customers, preferably mediated with their own merchant fleets. These latter-type port cities became conduits, accelerators, brokers and agents of *globalisation*: understood as the long-distance exchange of people, goods, money, technologies, ideas, beliefs and biota (animals, plants, microbes).

Of the port cities cited by Darwin (2020), most are parts of mainland coasts, entangled and embedded in commercial circulations that extend landward into a lush and rich interior: what eventually became Canada, USA, UK, China, India, Malaya, Persia, East and West Africa pivoted and privileged such ports as exporters of staples or processed goods (such as lumber from Canada; manufactures from China; steel, woven cotton and pottery from the UK; and, in the case of Africa, also slaves) and not just as *entrepôts*.

And yet, what would happen if the port city existed on a small island, to the extent that it evidently had no hinterland to speak of? Many *entrepôt* port cities were set up on small islands: small enough for easy defence; and obliging invaders to muster a fleet and wait for suitable weather conditions before any attack was even possible. Tyre (an island before it was turned into a peninsula by Alexander the Great), Venice, Hormuz, New Amsterdam (later New York), Montreal, Bombay, Penang, Singapore, Hong Kong, Lamu, Mombasa, Kilwa and Lagos were all of this type. Having *multiple* islands allowed for better quarantine provisions. Venice dedicated the islet of *Lazzaretto Vecchio* (old lazaret, in Italian) for this purpose (St Lazarus is the patron saint of lepers). Having deep, sheltered and ice-free harbours added to the value of the port city as an asset to its basic appeal as a spoil of conquest.

2 Origins

Historically, settlements were often established *away* from the coast and on high ground, even on islands, allowing both an easier visual patrol over longer distances; as well as a better defensive position, should an attack unfold. Kyoto, the old capital of Japan; Mdina, the old capital of Malta; Corte, the old capital of Corsica; Knossos, the old capital of Crete; Villa de Teguise, the old capital in Lanzarote; Betancuria, the oldest settlement of Fuerteventura; La Laguna, the old capital of Tenerife; Cinarli, the main town on the Turkish island of Gökçeada; Shuri, the capital of Ryukyu Kingdom/ Okinawa; and Anuradhapura, the old capital of Sri Lanka, are thus all located inland.

Large countries could afford to self-isolate, shut the door to trade and rely on domestic food and goods; or else carefully manage a limited international trade lest its cosmopolitan qualities crossed over and infected the local population: this was the case with Japan during the Edo period, that restricted European trade to the islets of Hirado and then Deshima (出島, or 'exit island') for over 200 years (e.g., Joshi and Kumar 2002). But the urge to trade and take to, and from, the sea has been quite irresistible. And so, in many cases, maritime trade grew in scale and depth, and new coastal island cities emerged, expanded and eclipsed the former (non-coastal) capital cities, assuming the title themselves. Valletta (Malta) since 1568; Ajaccio in Corsica; Colombo in Sri Lanka; Chania since 1847 and then Heraklion in Crete (Greece) since 1971; Santa Cruz in Tenerife (Canary Islands) since 1833; Arrecife in Lanzarote (also Canary Islands, Spain) since 1852; Naha in Okinawa after 1879; and Tokyo (Japan) since 1868.

For other islands that were invaded, wrested from their native population or else settled from scratch in the wake of the European Age of Discovery (and thus wholesale creations of globalisation), their main urban footprint was a port city from the very outset. Such coastal and port settlements—such as Port Louis, Mauritius; or Jamestown, St Helena—by definition, had to be located close to the shore, but this invariably made them more vulnerable to attack and surprise by pirates, corsairs or enemy navies. They thus had to be well defended, well-fortified and ideally under the patronage of a strong and reliable benefactor. Indeed, almost all capital cities on islands are coastal, embodiments of the cultural and commercial traffic that reflects maritime and commercial (rather than terrestrial and agrarian) economic systems (Murphy 2001). In contrast, those few capital cities on islands that are *not* coastal reflect economies that remain subsistence based (rather than export oriented); and/or which simply lack a suitable and sizable port. I can think of only four island jurisdictions which today do *not* have a coastal capital city; two being states; and the other two being territories: Madagascar (with Antananarivo as its inland capital), Cyprus (with Nicosia), Gozo (with Rabat, also known as Victoria) and Anguilla (with The Valley) (Baldacchino et al. 2024).

Warrington and Milne (2018: 191–192) describe as *entrepôts* those islands that are marketplaces, enjoying the advantage of location and centrality, while generating internally the conditions fostering their success: magnets for immigrant skilled labour, and investment finance; sites for entrepreneurial flair, repositories of a legal framework and a dispute resolution regime that collectively facilitate cross-border market transactions and innovation by minimising cost and risk. They single out Singapore, Guernsey and Jersey (the Channel Islands), Åland and Mauritius as examples of this 'political economy'; as well as Barbados, Ireland and Malta which have transitioned to this status. Unable to count on its own hinterland, an *entrepôt* must innovate tirelessly, enhance its competitiveness and protect its regional security. Indeed, such island port cities often depend for their safety and survival on geopolitical regimes that they do not control. They produce almost nothing themselves (Darwin 2020: 262).

Meanwhile, in many larger and non-island states, a different—even opposite—dynamic can be discerned. Less than one third of non-island states have a coastal capital city. Some such states are landlocked; so these have no sea-facing ports at all (although many would like to). Meanwhile, the largest states, typically with federal constitutions, have established or even shifted their capital 'inland', partly to better address the geographical sprawl of their territory, avoid the dominance of other large urban centres, and otherwise be closer to the centre of things. Think Abuja, Beijing, Brasilia, Canberra, Islamabad, Moscow, New Delhi, Ottawa or Pretoria. Even the archipelagic state of Indonesia, the world's fourth most populous country, has announced Nusantara as its new federal capital, to be located more centrally on Borneo. This move would replace Jakarta, the current capital, which lies on Java, a large island but on the country's geographic periphery. While located at the banks of rivers, capital cities like Berlin, Dhaka, Hanoi, Islamabad, Madrid, Mexico City, Paris, Tehran or Washington DC are not coastal (see Table 1). After Indonesia, Japan and the Philippines, the United Kingdom is the largest island country (by population) that has a port city, London, as its capital, reflecting its maritime imperial past and legacy.

Of course, the larger the country and its economy, the greater the share of its wealth that is likely to derive from *domestic* sources. This reduces the significance of import and export activity and, therefore, that of its ports. So, for example, the contribution of exports of goods and services to the Gross Domestic Product (GDP) of China, India and the United States was reported as 20, 21 and 11% respectively in 2021. Whereas for small states, including island states, the economic contribution of the exports of goods and services could equal *or even exceed* the country's annual GDP. The latter was the case for Djibouti, Ireland, Luxembourg, Malta, San Marino,

Table 1 Capital cities of the 193 UN member states

	Port capital city	Non-port capital city	Total
Island states	45 (98%)	2 (2%)	47 (100%)
Non-island states	44 (30%)	102 (70%)	146 (100%)
Total	*89*	*104*	*193*

Note Nauru is the only UN member state that does not have a capital city, but all its settlements are coastal. For countries with more than one capital city, such as The Netherlands, the data here refers to administrative capitals

Table 2 Economic contribution of exports to gross domestic product by % (2021)

Large non-island states	%	Small island states	%
China	19.9	Cyprus	89.4
India	21.5	Ireland	133.7
United States	10.9	Malta	166.4
Brazil	19.6	Seychelles	111.5
Russia	29.9	Singapore	184.3
South Africa	31.1		
World average (weighted by population) 29.1			

Seychelles and Singapore, also in 2021 (World Bank 2022). Indeed, trade dependency varies inversely with state size (Gerring and Veenendaal 2020, see Table 2).

3 Expansion

If it is a successful and thriving entrepôt, the port city—much as happens elsewhere—will tend to expand, typically frontally (via land reclamation, bridging or landfilling outlying islands) and/or reversely, by sprawling into its contiguous mainland, overrunning and incorporating its existing suburbs. What makes small island port cities rather distinct in this dynamic is the sheer scale of the impact and reach of their urbanisation. Capital cities and their suburbs on *small* states and territories—that is, those jurisdictions with resident populations of one million or less—are much more likely to contain disproportionate levels of the total population of that same state or territory than larger states. Within the Indo-Pacific region, the sovereign island state of Singapore is a 100% city state: the capital city accounts for the whole country. Bandar Seri Begawan, in Brunei; Koror City on Koror, capital of Palau until 2006; Metro-Honolulu within the US state of Hawaii; and Malé, capital of archipelagic Maldives: all embrace a majority of that respective island jurisdiction's population.

Already in the case of large islands, what was initially an archipelago has morphed into a cluster of connected islands that have additionally fused and spilled over adjoining mainlands: consider Montreal, in Quebec, Canada; or Stockholm, the capital of Sweden; or Bombay/Mumbai, in India (Grydehøj et al. 2015). On the smallest islands then, overdevelopment is an even more palpable threat. Many port city walls were dramatically breached or torn down, partially or totally, to allow for urban expansion and/or automobile access (Baldacchino 2014). Only strict zoning regulations and/or a modern appreciation for heritage or nature conservation may have prevented a widespread building spree, keeping the capital city growth in check, and maintaining a modicum of rurality in densely populated but small island settings (think Manhattan, with its iconic Central Park). Most of the small, non-island, European microstates—Andorra, Liechtenstein, Luxembourg, Monaco, San Marino, the Vatican—and territories (Gibraltar, Ceuta, Melilla)—as well as China's Special

Administrative Regions of Macao and Hong Kong—are effectively city states, all with high population densities. In *island* states, the situation is rendered more dire by the implications of geography: one cannot, for example, commute to and from work across the border on a daily basis. We thus come across the phenomenon of the *island jurisdiction as city state,* an urban agglomeration that effectively takes over much of the territory of the island state, as with Bahrain, Malta or Singapore (e.g., Sadoway 2013; Main et al. 2021); or islands (in the plural), in which case they are usually linked by bridges and/or causeways to each other: Bermuda or Macau, and partly the Maldives, would fit this latter description.

But, not quite. On small island jurisdictions, a few kilometres can suggest distance and isolation. While the ‘city as island’ escapes materially and conceptually from its fortified encasement, creeping out as both driver and consequence of globalisation in the process, the city also rolls over and materially embraces a scape that does not necessarily share its rubric of ideas and epistemologies (Baldacchino 2014). Even Bahrain, Malta and Singapore have a peri-urban zone and a ‘countryside’ that contrasts with the coastal capital city and its denser conurbation. For islands dabbling in tourism—and which island has not flirted with tourism?—the coast has been both blessed and maligned. Port cities on islands have seen their already stretched roles and infrastructures absorb additional functions or, where possible, expand to accommodate cruise ship terminals, which then disgorge thousands of visitors for short stays, adding to population pressure and traffic gridlock in the coastal capital (Navarro-Ruiz et al. 2020). On the other hand, and especially for those travellers arriving by plane, their destination is typically to less urban fringes, with their rustic settings and sandy beaches, revolutionising backwater settlements into tourist magnets, risking a Benidormisation of the coast and its associated trials: these include the volume of waste, alcoholism, pollution, and street violence (Baidal et al. 2013). Airports have now overtaken seaports as the preferred gateways for the entry and exit of persons on most islands, thus sidestepping the capital city; but this has not diminished the relatively towering status and size of island port cities as settlements. Indeed, airports on islands still tend to be located close to their capital cities; with *small* islands, this configuration is inevitable.

4 De-essentialising ‘the Island’

The actual personality and character of islands as liminal spaces, at once local and global, at once national and cosmopolitan, may depend crucially on the dynamics of their port (and often largest and/or capital) cities. It is by looking at what happens in the spaces of potential cultural exchange and brokerage of such locales—including working waterfronts, coastal tourism establishments and activities, markets and commercial transactions, religious rituals and celebrations—that one gets a better sense of whether and how islands, in spite (or because) of their smallness, can be ripe for human conviviality, multilingualism, tolerance and a peace based on commerce and its ensuing prosperity. As Montesquieu already argued in 1748:

> Commerce is a cure for the most destructive prejudices … [it] diffuses a knowledge of the manners of all nations … it is almost a general rule that, wherever we find agreeable manners, there commerce flourishes; and that wherever there is commerce, there we meet with agreeable manners.

For "crossroad islands" (Connell and King 1999), equivalent to the "entrepôts" of Warrington and Milne (2018), their openness to trade, ideas, cargo, investment and tourism is at once the main source of their vulnerability as well as of their strength. These islanders have embraced globalisation, for what it is worth, warts and all; and often without having any choice. They tend to be more tolerant of diversity and plurality. Of course, there is inequality, hierarchy, and tussles for power and influence. There is, as Foucault (1984: 9) reminds us, adventure and piracy. But islanders are bound to realise that autarky is not an option: they cannot block or reject foriegn trade. Their prosperity and quality of life are significantly premised on a committed disposition towards the outside.

In contrast, it is when island societies develop a narrow, exceptionalist and special definition of themselves— including pillarised notions of their own, internal, island communities in terms of race, ethnicity, nation—that they reduce their appeal as 'welcoming societies' and instead become proud, vocal and flag-bearing proponents of a pseudo-ethnicity, territoriality and local island greatness. Such island societies are nervous of globalisation; and consider immigration tantamount to invasion.

Such an understanding is called for because, as any islander would tell you, islands—even the smallest—are also riven by tensions, longstanding rivalries, antagonisms, fault lines and conflicts, which can turn violent (Austin 2000; Baldacchino 2012a; Richards 1982). Of course, because of their self-evident geographies, (small) islands are often reduced to caricatures: not just of isolation and insularity, but also of homogeneity and identity. Islanders stand as 'us', doggedly resilient against 'them': the equally essentialised Other. This is a trope that is romanced by external observers and scientists as much as by local raconteurs and popular history writers (e.g., Ratter 2018; Srebrnik 2004). Island citizens are expected to live together in harmony:

> [they] are likely to promote feelings of fellowship and a sense of community. When people live at a distance from the outside world, they share a feeling that they are, so to speak, alone in the world (Anckar, 2002: 386).

This impression of islands as naturally unitary spaces is reinforced by the fact that only ten of the world's inhabited islands are shared between more than one country (Baldacchino 2013).

Talk of an 'urban–rural divide' may be appealing and the dichotomy is beguiling and attractive; but it masks and downplays the very extensive dynamics that connect the two socio-ecological systems (e.g., Scott et al. 2007). The imbrication of the two in a small island setting is rendered more complex, because, on one hand, the port city is, by definition, less economically connected to its island 'hinterland': indeed, there may be no hinterland to speak of (Hintjens 1991: 38). This would explain why colonialism has been so deep, long, intimate and enduring for most small islands (Caldwell et al. 1980). On the other hand, the 'urban–rural' social and cultural connections are likely to be strong, as people navigate and move across the short distances between port city and village on a regular and circular basis.

5 Dealing with Others

The 'port-city and the rest' divide is vividly illustrated by how islanders, and their elites, deal with foreigners.

We start by acknowledging that the 'native-settler' alterity is a powerful marker of social dynamics in many islands (Houbert 1985, 2015). *Blow Ins, Come from Aways* and *Non-Belongers* are terms that apply to those who are deemed not to have been island born, and may thus lack access to certain rights and privileges, including the right to vote. Small island state societies with significant numbers of residents of Indian descent—Trinidad, Fiji, Mauritius—are riddled by ethnic markers which percolate also into political party mobilisation (with majorities in Trinidad and Mauritius; but minority status in Fiji) (e.g., Eriksen 1994; Srebrnik 2000). And while tourists are generally welcomed and feted as net contributors to economic wellbeing, the legendary island hospitality can change into closure, suspicion and indifference if visitors overstay (Baldacchino 2012b). Such 'insider–outsider' tensions can boil over when long-term immigrants and second home residents lobby to advance their own rights, developmental agendas and visions on local planners and policy makers; appealing, for example, for the conservation of tradition and legacy while the locals are more keen to catch up with the material trappings of modernity and exploit economic returns (e.g., Marshall 2009); or for newcomers to be open to homosexuality and same sex liaisons in the face of a more conservative and religious establishment (e.g., Vlcek 2013). When the four Atlantic provinces of Canada, including Prince Edward Island, first set up *immigration* strategies to address self-evident demographic shortfalls, what they really meant and hoped for initially was *repatriation*: to see those locals who had migrated 'come back home' (Fall 2008).

The arrival of long-term immigrants, usually on work visas, or of undocumented migrants, is an especially testing experience for small island societies. Even relatively moderate flows can soon assume dramatic proportions in small states and territories, fuelling xenophobia. When overwhelmed by such arrivals, an earlier disposition towards hospitality by the locals can morph into frustration, anger and support for right-wing political action.

For example, there were some 2,100 undocumented migrants in a welcome facility on the Italian island of Lampedusa (resident population: 6,500; land area: 20 square km) with enough beds to host 200, in July 2022 (Aljezeera 2022). In 2015, at the height of the European refugee crisis, some 21,000 asylum seekers were landed in Lampedusa. The island was a 2016 candidate for the Nobel peace prize for its solidarity toward asylum seekers; but has since became a stronghold of anti-immigration politics.

Lesvos (resident population: 80,000; land area: 1763 square km) is one of five Greek islands hosting EU-funded 'asylum hotspots' that are used as a buffer zone between Turkey and Europe. Since an EU-Turkey deal in 2016, thousands of asylum seekers have been trapped on these islands on their way to Europe. Both locals and refugees have been protesting for years against this policy and the creation of new refugee camps. Although political motives vary, their message is clear: Lesvos does not want to be Europe's 'refugee warehouse' (Vallianatu 2022).

It does *not* have to be this way. First, consider Malta (population: 526,000; land area: 316 square km) has seen over 100,000 immigrants settle in the country over just the last decade, reversing decades of a culture of emigration (Baldacchino 2017). Many chose to stay during Covid-19. The local economy (including tourism and construction) and the public health service depend crucially on foreign labour to function. Unemployment is at record lows and, while there is a racist streak (especially in relation to Moslem, sub-Saharan immigrants), the country has taken this development in its stride (Bradford and Clark 2014; Debono 2021). The only party with populist credentials is the radical right *Maltese Patriots Movement* (MPM) which is opposed to Islam, multiculturalism and immigration, but it is electorally irrelevant: it gained 0.3% of the vote in the 2019 elections to the European Parliament (Leusch 2019; Taggart and Pirro 2021).

Next, consider Singapore (population: 5.6 million; land area: 733 square km) where the most declared official religion is Buddhism, but by just 30% of the population. Already in the mid-nineteenth century, evolutionary biologist Alfred Wallace (1869) described Singapore's ethnic diversity: its English government, garrison and main merchants; its largely Chinese population; its Portuguese clerks and lesser merchants; its Arab shopkeepers; its Bengal grooms and washers; its Japanese sailors and domestic servants; plus traders from Celebes, Bali and beyond.

In both these countries, more so in Singapore than Malta, a clear democratic deficit lies with the growing proportion of residents who are not nationals and do not participate in the democratic process. Island port cities are gateways of migration and, if successful, can quickly lead to a situation where locals are outnumbered by immigrants. The condition holds the seeds of gentrification that may alienate local residents who, therefore, are understandably cautious in extending rights, such as residency and property ownership (Clark et al. 2007).

A third example is Gibraltar, the only Overseas Territory of the United Kingdom not to be an island, but which can be described as the most islanded of them all, given the longstanding claim to the territory by Spain, emboldened since Brexit, and with which Gibraltar shares its land border. Gibraltar's past reveals a complex racial mix reflecting immigration into an abandoned territory from various Mediterranean sources, notably from Genoa and other parts of Italy, Menorca, Portugal, Malta and Spain, as well as from Britain and North Africa (Archer 2006: 2).

6 Discussion

Port cities face the sea; and technologies of maritime transport—from small skiffs and Filipino batels to container vessels—are their livelihood, and their tools of choice to engage the world (Santarita 2021). As Foucault (1984: 9) argues, boats have not only been, for our civilization, from the sixteenth century onwards, the great instrument of economic development. They have also been the greatest reserve of the imagination. The ship is the heterotopia [or counter-place] *par excellence.* In civilizations without boats, dreams dry up, espionage takes the place of adventure, and the police take the place of pirates. If the capital city stands for the rule-based superego, the port city stands for the rebellious id.

Because of their self-evident geographies, (small) islands are often reduced to caricatures of isolation and insularity; this approach has been savagely criticized by an alternative and just as essentialized representation of islands as spaces of flows and connectivity. But such binaries do not explain anything. The actual personality and character of islands as liminal spaces, spaces 'in between' (Baldacchino 2008), at once local and global, at once national and cosmopolitan, depends on the dynamics of their port (and often capital) cities. One needs to look closely at (1) *what happens* in the spaces of potential cultural exchange and brokerage in such locales—including working waterfronts, coastal tourism establishments, religious rituals and celebrations, (2) *the speed or otherwise by which that exchange happens*, and (3) whether islanders have *any control* on such exchanges and mobilities. It is thus that one gets a better sense of whether and how islands, in spite (or because) of their smallness, can be ripe for human conviviality, multilingualism, tolerance and a peace based on prosperity. They hold the promise for a global, and "progressive sense of place" (Massey 2008). In contrast, it is when island societies develop a narrow, exceptionalist and special definition of themselves—including pillarised notions of their own, internal island communities in terms of race, ethnic stock and national identity—that they reduce their appeal as 'welcoming societies' and instead become the strategic bulwarks, flanks and edges of hard borders. Being overwhelmed by the rapid arrival of large numbers of migrants, and over whose movement and possible settlement they have little control, is likely to exacerbate the islanders' siege mentality (e.g., Dodds 2020).

Islands feature prominently as 'mobilityscapes' (King 2009: 62–63). They served, and still serve, three main dynamics: as the *origins* of mobility (with islanders being keen transnationals who want to leave, 'see the world', and sometimes return); as the *destinations* of mobility (for tourism, travel, escapism, 'dropping out', 'moving in and moving on', and so on); and as locales for the *circular* flow of mobility implied by trade, commerce and other transactions.

The treatment of Jews as commercially significant—but historically ethnically contemptible—ethnic minorities in port cities is instructive here (Cesarani 2014). Are we to come across experiences more like William Shakespeare's Shylock in *A Merchant of Venice*, cast as a bloodthirsty moneylender, a tragic figure broken by a ceaseless persecution? Or more like Sir Walter Scott's Isaac of York in *Ivanhoe*, a more complex character, capable of responding with fortitude to threats or dangers, and of acting toward others—Christians as well as Jews—with genuine human sympathy, magnanimity and generosity?

7 Conclusion

Thus straddled between here and there, land and sea, in and out, the ebb and flow of goods, ships and people, port cities have the capacity to act uniquely "as crucibles of antinomical creation" (Murphy 2001: 22).

Abulafia (2011) goes to great lengths to explain the complex dynamics across the Mediterranean littoral that have prevailed across the ages. Although this 'great sea' has been the location of so many collisions (East versus West; Christendom versus Islam; Europe versus Africa …), the imperative of trade and commerce has kept port cities throbbing and operating. Venice—a rare case of an archipelago that itself became an empire—thrived by not bothering too much with religious dogmas and divides. The decline then started at the end of the 1400 s, as new worlds and lucrative routes were discovered beyond Europe. It was, however, the emergence of state-sponsored nationalism in the late 19th and early twentieth century, then cemented by two world wars, that saw a decline in the openness of trade and migration in the region: Smyrna, now Izmir, is a case in point, becoming a posterchild victim of nationalist ambition (Abulafia 2011; Darwin 2020; Chap. 9). And, in the contemporary moment, the border zone is a zone of state control, surveillance and enforcement; and increasingly since 1994, with coasts and islands becoming the pivots of large exclusive economic zones, thanks to the ratification of the UN Convention of the Law of the Sea (Baldacchino 2022). The relative freedom of passage of the past has taken on a new character. Some travellers—political and business elites; high net worth individuals; holders of 'golden passports'—can navigate the border relatively easily; in contrast, many common mortals must secure valid passports and visas; while still others can be denied entry and face detention and/or forced repatriation (Baldacchino 2018). Meanwhile, a legacy of the pre-statist past lingers on, in the provision of free ports and free trade zones (Gilmour 2022).

Gerring et al. (2022: 313–314) collect the evidence for arguing that islands around the world are also bulwarks of democracy. First, being irrevocably exposed to the sea increases the propensity towards democracy, since merchants and professionals engaged in maritime affairs have a strong incentive to support property rights and the rule of law—rather than dynasty or military might (Gerring et al. 2022: 50). Second, islands offered appealing ports of call and settlement to Europeans which—especially in the case of *small* islands—soon translated into a deep and enduring colonial presence: such a colonial experience was probably most transformative on and to islands and islanders, infusing them with the experience of electoral politics and semi-autonomous government which, in many cases, then translated into democratic practices, within a sovereignty or sub-national jurisdictional status arrangement. Third, most islands depend on international trade, finance or tourism for a large share of their local income; this may encourage a more open attitude towards democracy. Fourth, being geographically isolated, island states and territories may be less militaristic because their sovereignty is more secure than similarly-sized, land-based states and because, for them, expansionist policies are harder to pursue. Indeed, as far as is known, there have only been six 'island empires' throughout the world in recorded history: those of Minos (Crete) from 3000 to 1100BC, Tui Manu'a of Samoa and then Tu'i Tonga, around 1000 CE; Venice (circa 1000–1797), Britain (eighteenth-twentieth century) and Japan (1868–1945) (Warrington and Milne 2018). Gerring et al. (2022) add a fifth explanation: that island living may foster a greater sense of social harmony and national community (Anckar and Anckar 1995; Congdon Fors 2014; Royle 2002). But, as argued above, this latter claim should be seriously disputed.

In the past few decades, the study of islands *qua* islands has advanced considerably: the condition of *islandness* associated with these discrete pieces of land surrounded by water is now also recognised as an intermediate variable in various social and natural science texts, from archaeology to zoology. To the more obvious features of island studies—distance from mainland, size and topography, archipelago city and location, the role of the sea—must be added the availability or otherwise of a suitable harbour. This island attribute has been recognized at least 500 years: the port features prominently in the woodcut map of the 1516 first edition of Thomas More's *Utopia* (see Fig. 1). Trajectories of development, population and diversity, perhaps even the very fabric of communities and their ability to live together, have been driven and impacted, though not determined, by the presence or absence of such a valuable natural asset.

Fig. 1 The image of the island of Utopia, with the port in the foreground. (*Source* More 1516. Not subject to copyright)

Acknowledgements and Disclaimer This chapter is based on a keynote address delivered at an international conference held at Osaka Metropolitan University, Japan, in April 2023. I am grateful to Prof. Takashi Yamazaki (and his sponsors) for inviting me to this exciting conference and for his subsequent critically constructive comments on an earlier draft of this chapter. Parts of the argument for this chapter were developed in Baldacchino (2014) and in Baldacchino et al. (2024). The usual disclaimers apply. As the chapter's sole author, I declare no conflict of interest in its drafting.

References

Abulafia D (2011) The great sea: a human history of the Mediterranean. Oxford University Press, Oxford

Aljezeera (2022) Italy relocates refugees after Lampedusa centre overwhelmed. https://www.aljazeera.com/news/2022/7/10/italy-relocates-refugees-after-lampedusa-center-overwhelmed#:~:text=%E2%80%9CThere%20are%202%2C100%20people%20packed,are%20the%20ones%20who%20survived.%E2%80%9D Accessed 24 Sep 2024

Anckar D (2002) Why are small island states democracies? Round Tab Commonwealth J Int Aff 91(365):375–390. https://doi.org/10.1080/0035853022000010344

Anckar D, Anckar C (1995) Size, insularity and democracy. Scand Polit Stud 18(2):211–229. https://tidsskrift.dk/scandinavian_political_studies/article/view/32846 Accessed 24 Sep 2024

Archer EG (2006) Gibraltar, identity and empire. Routledge, London
Austin D (2000) Contested islands. Round Tab Commonwealth J Int Aff 89(353):59–63. https://doi.org/10.1080/750459456
Baidal JAI, Sánchez IR, Rebollo JFV (2013) The evolution of mass tourism destinations: new approaches beyond deterministic models in Benidorm (Spain). Tour Manage 34:184–195. https://doi.org/10.1016/j.tourman.2012.04.009
Baldacchino G (2012a) Islands and despots. Commonwealth Comp Polit 50(1):103–120. https://doi.org/10.1080/14662043.2012.642119
Baldacchino G (2012b) Come visit, but don't overstay: critiquing a welcoming society. Int J Cult Tour Hosp Res 6(2):145–153. https://doi.org/10.1108/17506181211233072
Baldacchino G (ed) (2013) The political economy of divided islands: unified geographies, multiple polities. Palgrave Macmillan, London
Baldacchino G (2022) UNCLOS at 40: small island developing states and new forays into positive sovereignty. Round Tab Commonwealth J Int Aff 111(5):598–611. https://doi.org/10.1080/00358533.2022.2129635
Baldacchino G, Oroz T, Čuka A (2024) Capital Zagreb versus coastal split: enduring tensions between capital cities and port second cities. Sociol I Prost 62(2):235–256. https://hrcak.srce.hr/321014 Accessed 30 November 2024.
Baldacchino G (2008) Islands in between: Martín García and other geopolitical flashpoints. Island Stud J 3(2):211–224. https://doi.org/10.24043/isj.223
Baldacchino G (2014) Capital and port cities on small islands sallying forth beyond their walls: a Mediterranean exercise. J Medit Stud 23(2):137–151. https://www.um.edu.mt/library/oar//handle/123456789/16517 Accessed 24 Sep 2024
Baldacchino G (2015) Lingering colonial outlier yet miniature continent: notes from the Sicilian archipelago. Shima Int J Res Island Cult 9(2):89–102. https://www.um.edu.mt/library/oar//handle/123456789/16556 Accessed 30 Nov 2024
Baldacchino G (2017) Malta and the new normal: the labour importing economy. In: Biennial report of the centre for labour studies 2015–2016. Centre for Labour Studies, University of Malta, Malta, pp 11–18. https://www.um.edu.mt/library/oar/bitstream/123456789/40954/1/Biennial_report_2015_2016.pdf Accessed 24 Sep 2024
Baldacchino G (2018) Connectivity, mobility and island life: parallel narratives from Malta and Lesvos. Symp Melit 14:7–17. https://www.um.edu.mt/library/oar//handle/123456789/30006 Accessed 24 Sep 2024
Bradford S, Clark M (2014) Strangers on the shore: Sub-Saharan African "irregular" migrants in Malta. J Immigr Refug Stud 12(1):9–26. https://doi.org/10.1080/15562948.2013.877703
Caldwell JC, Harrison GE, Quiggin P (1980) The demography of micro-states. World Dev 8(12):953–968
Cesarani D (2014) Port Jews. Routledge, London
Clark E, Hedin K, Lundholm E, Malmberg G (2007) Island gentrification and space wars. In: Baldacchino G (ed) A world of islands: an island studies reader. Institute of Island Studies, University of Prince Edward Island and Agenda Academic, Canada and Malta, pp 483–512
Congdon Fors H (2014) Do island states have better institutions? J Comp Econ 42(1):34–60. http://www.sciencedirect.com/science/article/pii/S0147596713000930 Accessed 24 Sep 2024
Connell J, King R (1999) Island migration in a changing world. In: King R, Connell J (eds) Small worlds, global lives: Islands and migration. Pinter, London, pp 1–26
Darwin J (2020) Unlocking the world: port cities and globalisation in the age of steam, 1830–1930. Penguin, London
Debono M (2021) Migrants and the challenge of decent work in Malta. E-Rev Int Protección Soc 6:272–293
Dodds K (2020) God save the Falklands: postcolonial geographies of the Falklands/Malvinas. In: Edmond R, Smith V (eds) Islands in history and representation. Routledge, London, pp 177–189
Eriksen TH (1994) Nationalism, mauritian style: cultural unity and ethnic diversity. Comp Stud Soc Hist 36(3):549–574. https://doi.org/10.1017/S001041750001923X

Fall C (2008) Immigration, repatriation and retention: population strategies on Prince Edward Island and comparable jurisdictions. In: Baldacchino G, Stuart K (eds) Pulling strings: policy insights for Prince Edward Island from other sub-national island jurisdictions. Island Studies Press, Charlottetown, pp 53–71. https://www.um.edu.mt/library/oar//handle/123456789/20191 Accessed 30 November 2024.

Foucault M (1984/1967) Of other spaces: Utopias and heterotopias. From Architecture/Mouvement/Continuité. Translated from the original French by Jay Miskowiec. https://web.mit.edu/allanmc/www/foucault1.pdf Accessed 24 Sep 2024

Gerring J, Veenendaal W (2020) Population and politics: the impact of scale. Cambridge University Press, Cambridge

Gerring J, Apfeld B, Wig T, Tollefsen AF (2022) The deep roots of modern democracy: geography and the diffusion of political institutions. Cambridge University Press, Cambridge

Gilmour PM (2022) Freeports: innovative trading hubs or centres for money laundering and tax evasion? J Money Launder Control 25(1):63–71. https://doi.org/10.1108/JMLC-01-2021-0002

Grydehøj A, Pinya XB, Cooke G, Doratlı N, Elewa A, Kelman I, Pugh J, Schick L, Swaminathan R (2015) Returning from the horizon: introducing urban island studies. Urban Island Stud 1(1):1–19. https://doi.org/10.20958/uis.2015.1

Hintjens HM (1991) France in the Caribbean. In: Sutton PK (ed) Europe and the Caribbean. Macmillan, London, pp 37–70

Houbert J (1985) Settlers and natives in decolonization: the case of New Caledonia. Round Tab Commonwealth J Int Aff 74(295):217–229

Houbert J (2015) The West in the geopolitics of the Indian Ocean and India. In: Rumley D, Chaturvedi S (eds) Geopolitical orientations, regionalism and security in the Indian Ocean. Routledge, London, pp 56–83

Joshi JS, Kumar R (2002) The Dutch physicians at Dejima or Deshima and the rise of Western medicine in Japan. Proc Indian Hist Cong 63:1062–1072. Indian History Congress. https://www.jstor.org/stable/pdf/44158176.pdf?casa_token=xuQtklsCBUMAAAAA:NDlz1ibgiTeTIODkrjlPw4wuOht8XfS9zJQlyWwdHzWkCgqhP4cBkxMgoZ6VlTL2S73aAzLvrlGX9xPNHveaYbpC-T-hN8AL2MCK0WPkOOwnURtFww Accessed 24 Sep 2024

King R (2009) Geography, islands and migration in an era of global mobility. Island Stud J 4(1):53–84. https://doi.org/10.24043/isj.228

Leusch C (2019) Maltese exceptionalism in the populist trend. MaRBLe, 1. https://doi.org/10.26481/marble.2019.v1.752

Main G, Schembri J, Speake J, Gauci R, Chester D (2021) The city-island-state, wounding cascade, and multi-level vulnerability explored through the lens of Malta. Area 53(2):272–282. https://doi.org/10.1111/area.12709

Marshall J (2009) Tides of change on Grand Manan island: culture and belonging in a fishing community. McGill-Queen's University Press, Kingston Ontario, Canada

Massey D (2008) A global sense of place. The cultural geography reader. Routledge, London, pp 269–275

Montesquieu (1748) Spirit of Laws. Book 20. Chapter 4. https://press-pubs.uchicago.edu/founders/documents/v1ch4s2.html#:~:text=Commerce%20is%20a%20cure%20for,now%20less%20savage%20than%20formerly Accessed 24 Sep 2024

More T (1516) Libellus vere aureus, nec minus salutaris quam festivus, de optimo rei publicae statu deque nova insula Utopia

Murphy P (2001) Marine reason. *Thesis Eleven 67*, 11–37. https://journals.sagepub.com/doi/pdf/https://doi.org/10.1177/0725513601067000003?casa_token=9lpASm2kHtEAAAAA:c08JfBsold8VFSSuIGtt4dm6cZJWFOp94gjfj0hYWt-J5kzfquW0RPYVmDfcMBMBqyGsj4BKWg Accessed 24 Sep 2024

Navarro-Ruiz S, Casado-Díaz AB, Ivars-Baidal J (2020) Cruise tourism: the role of shore excursions in the overcrowding of cities. Int J Tour Cities 6(1):197–214. https://doi.org/10.1108/IJTC-04-2018-0029

Ratter BM (2018) Island vulnerability and resilience. Geography of small islands: outposts of globalisation. Springer, Berlin, pp 173–199

Richards J (1982) Politics in small independent communities: conflict or consensus? J Commonwealth Comp Polit 20(2):155–171

Royle SA (2002) Geography of islands: small island insularity. Routledge, London

Sadoway D (2013) How are ICTs transforming civic space in Singapore? Changing civic–cyber environmentalism in the island city-state. J Creat Commun 8(1):107–138. https://doi.org/10.1177/0973258613512576

Santarita J (2021) Plying the straits: batel mobilities in central Philippines. University of the Philippines Diliman, Metro Manila

Scott A, Gilbert A, Gelan A (2007) The urban-rural divide: myth or reality? Macaulay Institute, Aberdeen

Srebrnik HF (2000) Can an ethnically-based civil society succeed? The case of Mauritius. J Contemp Afr Stud 18(1):7–20. https://doi.org/10.1080/025890000111940

Srebrnik HF (2004) Small island nations and democratic values. World Dev 32(2):329–341. https://doi.org/10.1016/j.worlddev.2003.08.005

Taggart P, Pirro AL (2021) European populism before the pandemic: ideology, Euroscepticism, electoral performance, and government participation of 63 parties in 30 countries. Italian Polit Sci Rev/Riv 51(3):281–304. https://doi.org/10.1017/ipo.2021.13

Vallianatu AI (2022) Lesvos: how EU asylum policy created a refugee prison in paradise. Chatham House. https://www.chathamhouse.org/2022/07/lesvos-how-eu-asylum-policy-created-refugee-prison-paradise#:~:text=Lesvos%20is%20one%20of%20five,on%20their%20way%20to%20Europe Accessed 24 Sep 2024

Vlcek W (2013) Crafting human rights in a constitution: gay rights in the Cayman Islands and the limits to global norm diffusion. Glob Const 2(3):345–372. https://doi.org/10.1017/S2045381713000142

Wallace AR (1869) The Malay archipelago. The land of the orang-utan, and the bird of paradise. A narrative of travel, with studies of man and nature. Macmillan, Oxford

Warrington E, Milne D (2018) Island governance. In: Baldacchino G (ed) The Routledge international handbook of island studies: a world of islands. Routledge, London, pp 173–201

World Bank (2022) Exports of goods and services as % of GDP. https://data.worldbank.org/indicator/NE.EXP.GNFS.ZS?end=2021&locations=CN&start=1960 Accessed 24 Sep 2024

Construction of the Artificial Island of Hulhumalé as a Climate Change Adaptation Measure in the Maldives

Akiko Sakamoto, Aishath Laila, Mariyam Areesha Ahmed, Daisuke Sasaki, Miko Maekawa, Ahmed Aslam, and Mikiyasu Nakayama

Abstract The Maldives, a low-lying atoll nation in the Indian Ocean, highly vulnerable to climate change impacts such as rising sea levels, has initiated the construction of the artificial island of Hulhumalé as a climate change adaptation measure. The artificial island of Hulhumalé is being built on a reclaimed land area of 400 ha. It is essentially constructed as an expansion of Malé, the capital city of the Maldives. Hulhumalé is also designed to be resilient to the effects of sea-level rise, with an average elevation of 2 m. It is designed as a fusion of nature and urban life, meant to attract many domestic migrants to relocate there. This chapter evaluates the socio-economic and environmental aspects of Hulhumalé's development by analyzing residents' perspectives on living conditions and migration patterns. Additionally, it explores the role of the Housing Development Corporation (HDC) in orchestrating the island's expansion and the challenges associated with financing, housing affordability, and environmental sustainability. This chapter also underscores the need for comprehensive approaches to climate change adaptation, highlighting the importance of sustainable infrastructure development, inclusive housing policies, and mental health considerations in atoll nations' resilience-building efforts. Hulhumalé has the potential to serve as a model for other atoll countries facing similar challenges due to climate change. However, and as with other artificial island projects, further surveys and discussions are necessary on such issues like the large construction cost and international debt, housing cost and rents for vulnerable people, mental health issues of migrants, and the overall environmental impact of the construction.

A. Sakamoto (✉) · M. Nakayama
Global Infrastructure Fund Research Foundation Japan, Tokyo, Japan
e-mail: a.sakamoto@gif.or.jp

A. Laila · M. A. Ahmed · A. Aslam
Housing Development Corporation Limited, Hulhumalé, Maldives

D. Sasaki
International Research Institute of Disaster Science, Tohoku University, Sendai, Japan

M. Maekawa
The Sasakawa Peace Foundation, Tokyo, Japan

T. Yamazaki and G. Baldacchino (eds.), *Relational Geographies of Islands in the Indo-Pacific*, International Perspectives in Geography 23,
https://doi.org/10.1007/978-981-96-7155-7_2

Keywords Maldives · Climate change adaptation · Right to Stay · Artificial island · Internal migration

1 Introduction

1.1 Coping with Climate Change by Small Island States

There are four atoll countries in the world, all being exclusively formed from coral reefs: Kiribati, Maldives, the Marshall Islands, and Tuvalu. The average natural land elevation of an atoll is extremely low: 2.1 m for the Marshall Islands and Kiribati; and just 1.5 m for the Maldives and Tuvalu. The low altitude of these countries makes them particularly vulnerable to specific effects of climate change, such as rising sea levels, storm surges, king tides and other, extreme weather events. As sea levels rise, these countries will be increasingly at risk of flooding and may eventually become uninhabitable (Barnett and Adger 2003; Betzold 2015). With the disappearance of the sum total of their material territory beneath the surface of the ocean, the very existence of these small, sovereign states can be brought into question at international law (e.g., Yamamoto and Esteban 2010; Hioureas and Camprubí 2020).

According to the IPCC (2019), global mean sea level is likely to rise between 0.29 and 1.1 m by the end of this century. The lower boundary of this range is based on the assumption that global warming is limited to an increase of 1.5 °C above pre-industrial levels, while the upper bound is based on the assumption that the increase in warming will reach 2 °C. Figure 1 shows the projected sea level rise for Malé, the capital of the Maldives, according to NASA's Sea Level Projection Tool. SSP1-1.9 represents one of the most stringent climate action scenarios, aiming to limit global temperature rise to below 1.5 °C by 2100, in line with the Paris Agreement. SSP1-2.6 envisions relatively successful climate mitigation efforts, aiming to keep the temperature rise below 2 °C. The projected sea level rise for Malé is 0.43 m under SSP1-1.9 and 0.49 m under SSP1-2.6.

In the United Nations Framework Convention on Climate Change (UNFCCC), there has been a shift towards emphasizing adaptation, without abandoning efforts to mitigate climate change (Kuyper et al. 2018). Both aspects are recognized as commendable, and technological solutions and other issues are being addressed.

In discussions regarding climate change, migration is often mentioned as an adaptive measure to address challenges such as rising sea levels, providing an additional option.

Nakayama et al. (2022) examined the pros and cons of the following four climate change adaptation options that are deemed realistic for the Government of the Republic of the Marshall Islands to contemplate: (1) migration to the developed world; (2) migration to other island states; (3) land reclamation and raising; and (4) development of floating platforms. The Government of the Maldives has invested seriously in Option 3, as this chapter will illustrate.

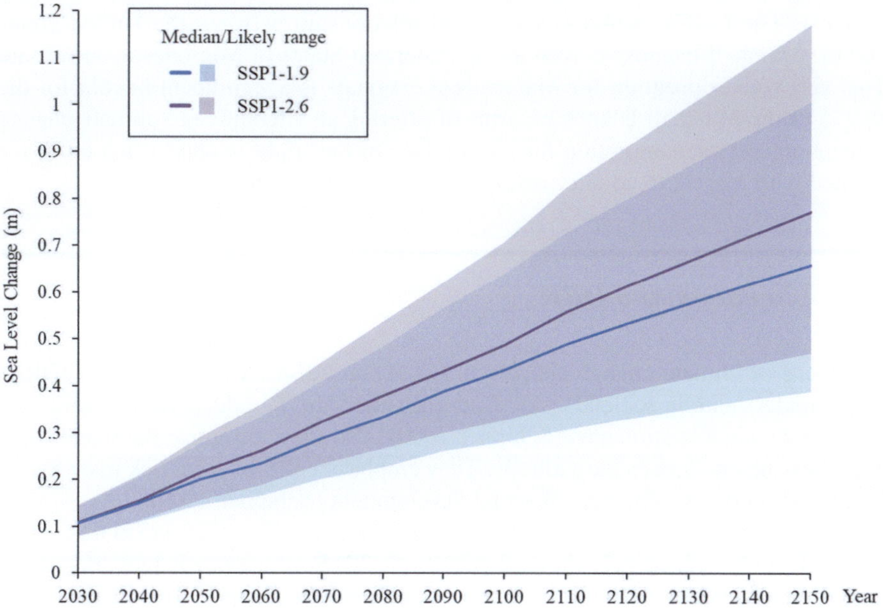

Fig. 1 Projected sea level rise under different SSP scenarios for Malé (*Source* IPCC sea level projection tool by NASA https://sealevel.nasa.gov/ipcc-ar6-sea-level-projection-tool, Accessed 26 September 2024)

Atoll nations face inevitable sea-level rise due to climate change, necessitating adaptation measures to protect their people. One route to take for dealing with this issue includes relocating to higher ground; however, this method is impractical for nations without elevated areas. Building seawalls is another option, yet their effectiveness diminishes as sea levels continue to rise over time. Migration abroad presents challenges: not only due to cost issues, but also because not all sovereign states accept climate migrants. Additionally, adapting to a new culture poses difficulties. Overall, while these measures offer short-term solutions, they may not suffice in the long run. Atoll nations must carefully consider their options for climate change adaptation.

Nakayama et al. (2022) argued that, in terms of possible climate change adaptation options for the Marshall Islands, "migration to developed countries" is the best choice from a post-migration livelihood perspective, followed by "development of floating platforms," "land reclamation and uplift," and "migration to other island nations." It was also found that "migration to the developed world" offered the most societal change to immigrants, while the "land reclamation and raising" alternative provided the least change.

It should be noted that the option of "migration to the developed world" is already readily available to nationals of the Marshall Islands, while it is not the case for nationals of the other three atoll countries. Under the terms of the Compact of Free Association (COFA)—signed in 1986 (Department of State 2003)—Marshallese nationals are not required to obtain a visa to enter the United States for tourism or business. The COFA establishes a special relationship between the United States and the Marshall Islands, as well as the Federated States of Micronesia, and Palau. This visa waiver program for Marshallese nationals is a significant benefit for the people of the Marshall Islands in terms of offering an alternative to climate change adaptation; yet such an option flies in the face of the 'right to stay' often taken for granted with respect to country residents: not everyone wants to migrate.

1.2 Adaptation and DRR

The focus of climate change adaptation (CCA) and disaster risk reduction (DRR) lies in understanding societal risks. Both measures aim to reduce vulnerability and develop sustainable strategies (Schipper 2009). Yet, they recognize the inadequacy of addressing challenges individually. They emphasize the inherent risk in everyday life, stressing the significance of social development (Schipper 2009).

The Sendai Framework for Disaster Risk Reduction 2015–2030 (SFDRR) aims to significantly reduce risk and loss (UNISDR 2015). It acknowledges in-country migration as a response to climate-induced hazards (Guadagno 2016; Djalante and Lassa 2019). The International Organization for Migration (IOM) recognizes SFDRR as the first global agreement to address mobility and displacement (IOM 2022).

SFDRR highlights the importance of managing migration in DRR, emphasizing community resilience while also minimizing displacement risks. Integrating migration into DRR strategies helps countries address climate-induced migration challenges and promote sustainable development for displaced populations and host communities.

1.3 Maintain the "Right to Stay"

The concept of "Right to Stay" has evolved over time. It was initially discussed in the context of forcibly relocating people due to large-scale infrastructure development. Recently, it has gained prominence within climate change adaptation discussions. According to a 2021 White House report, permanent planned relocation should be a last resort (White House 2021: 23). Instead, efforts should focus on enabling people to stay safely in their current homes by investing in disaster risk reduction and local adaptation measures (Asian Coalition for Housing Rights (ACHR) 2004).

The term "Right to Stay" refers to allowing individuals—who may eventually face permanent relocation—to remain in their traditional homes for as long as possible. It may serve as an interim measure rather than a final solution. However, atoll countries can ensure long-term 'Right to Stay' by commissioning suitable and climate change-proof infrastructure development. For instance, the Marshall Islands government explored reclaiming and raising land in 2019, emphasizing their people's commitment to stay in their country regardless of the cost (RNZ 2019). Similarly, Kiribati's

President Taneti Maamau advocated for elevating Kiribati's land in order to combat rising seas during the 2022 United Nations climate talks (Milman 2022).

1.4 Literature Review

Climate change has emerged as a critical global challenge, and one that particularly impacts low-lying island nations. This select literature review delves into key studies that address the implications of climate change on atoll countries, focusing on themes such as sea-level rise, extreme weather events, adaptation, and international legal considerations.

In examining the impact of climate change on Small Island Developing States (SIDS), Betzold (2015) provides valuable insights into adaptation strategies. SIDS, being among the first and worst affected by climate change, are viewed as "early adaptors," with the potential to offer key lessons for broader adaptation efforts. The paper focuses on migration as part of adaptation, emphasizing both its historical relevance in island life and its potential to fund adaptation measures through remittances. However, Betzold notes the challenges that hinder in-situ adaptation, categorizing them into perceptions and awareness, institutional issues, and resource constraints. The paper argues for overcoming these barriers by providing information and resources at the local level, enabling island communities to make informed decisions and adapt successfully. This text underscores the importance of recognizing the dynamic and multifaceted nature of adaptation in the context of SIDS facing the impacts of climate change.

Hioureas and Camprubi's (2020) exploration of the Montevideo Convention addresses the preservation of low-lying small island states facing rising sea levels. The article surveys various theories, including maritime delimitation agreements, land purchase, cession of territory, and the controversial concept of constructing new "territory." The authors analyze legal and political implications, considering the feasibility of each solution and existing state practices. This text provides a comprehensive understanding of the legal challenges that are associated with the preservation of existing territory and entitlements under the jurisdiction of low-lying island and archipelago states. The consideration of maritime boundaries and potential solutions like land purchase or cession adds depth to the discussion on the legal implications of climate-induced challenges for these states.

Climate change-induced threats to atoll countries, as outlined by Barnett and Adger (2003), constitute a perilous challenge to the long-term habitability of these regions. The authors emphasize the potential dangers of sea-level rise, sea-surface warming, and extreme weather events, framing these changes as risks that could compromise national sovereignty. The study introduces novel challenges for both climate change research and policy. The research challenge lies in identifying critical thresholds that could lead to the collapse of atoll social-ecological systems. Barnett and Adger (2003) argue that these thresholds are not only ecologically driven, but also behaviorally influenced by expectations in resource management. From a policy perspective, the authors call for the recognition of the unique vulnerability of atoll countries within the UNFCCC framework. Operationalizing international norms of justice, sovereignty, and both human and national security are all identified as crucial components in mitigating the impacts of climate change on atoll nations.

As for ways and means of addressing the issue, Schipper (2009) emphasizes the need for stronger links between CCA and DDR, recognizing the common goal of reducing vulnerability and creating sustainable long-term strategies. The implicit acknowledgment that risk is part of everyday life underscores the vital role of social development in both adaptation and disaster risk reduction. The potential convergence of these two tracks prompts questions about desirability and expected outcomes, helping to pave the way for future research and policy development.

Stojanov et al. (2017) explore the concerns of Maldivians regarding rising sea levels and immediate non-climate-related issues. This study highlights the urgency of addressing issues such as lack of housing, overcrowding, freshwater scarcity, and inadequate waste management. While sea-level rise is a concern for the future, slow-onset hazards like coastal erosion and changes in monsoon patterns are already affecting local livelihoods. The consideration of migration as an adaptation strategy, especially given the Maldives' location and attitudes toward climate change, adds a practical dimension to the discussion on climate-induced migration.

Wille (2023) addresses the depletion of coral rock in the Maldives due to construction industry demand. Current practices harm reefs, affecting both biodiversity and fish abundance. Recovery times may exceed 50 years, which raises concerns about long-term ecological impact. The study suggests alternatives like concrete blocks but also emphasizes the need for quality control.

Yamamoto & Esteban (2010) presciently delves into the potential physical disappearance of low-lying island states due to sea-level rise. The examination of likely scenarios and the possibility of these states claiming an Exclusive Economic Zone—even without any territorial presence above the surface of the sea—raises crucial questions about the future status of island states. The discussion of a "government-in-exile" adds a unique perspective, contemplating the retention of identity while awaiting a potential re-emergence of the island. This paper contributes to the broader discourse on the geopolitical implications of sea-level rise and the potential redefinition of statehood in the face of territorial losses.

These works collectively contribute to a comprehensive understanding of the multifaceted challenges posed by climate change to atoll countries (the Maldives in particular); they offer pertinent insights into ecological, societal, legal, and governance dimensions. The collective literature underscores the urgency of addressing these challenges through a combination of interdisciplinary research and effective international policies.

After reviewing these articles, it is felt that the option of artificial island development has not been adequately addressed as a measure of climate change adaptation for atoll countries. Conducting a study on Hulhumalé, an already developed and populated artificial island in the Maldives, would provide more insights into the feasibility of this option.

2 Aim and Study Area of This Study

2.1 Objectives of This Study

Authors, particularly Nakayama, Sakamoto, and Sasaki, have studied and published on the issue of internal and external migration of atoll countries. Nakayama et al. (2022) examined four options for the Marshall Islands to cope with sea level rise: migration to developed countries or other islands, land reclamation, and floating platforms. Using analytical methods, they found that migration to developed countries was the best option in terms of monetary benefits, while it tended to cause social problems. Sakamoto et al. (2022) focus on the artificial island of Hulhumalé in the Maldives as a solution to sea level rise. Current migration to Hulhumalé is driven by employment and education, not climate concerns. The study recommends job placement programs and IT-based social networks to facilitate smooth migration and resettlement. Sasaki et al. (2024) examined factors that influence migrant satisfaction in Hulhumalé. This survey is presented in more detail in 2.3.1.

Based on these previous studies, the authors concluded that an in-depth study of the livelihoods of migrants in Hulhumalé is desirable.

The aim of this study is to determine the gains and losses of the construction of the artificial island of Hulhumalé by the Maldives and the migration of its citizens to the area. The construction of artificial islands is considered to be one of the most promising options for climate change mitigation measures. Findings from this study should provide useful clues for other atoll countries in both developing and implementing climate change adaptation measures.

2.2 Desire for Better Living Conditions Among Residents of the Greater Malé Region

This chapter reviews the inspiration, conception, and operationalization of an ambitious plan that is currently underway in the Maldives (Fig. 2): to build an artificial island that is relatively safe from sea-level rise with the intention to serve as the new home for a significant proportion of the country's population. With the required expertise and loans in place, a new island with an area of 4 km^2 has risen in North Malé Atoll.

Malé, serving as the capital island, has been the nucleus of all services in the Maldives. With commendable healthcare and educational facilities, Malé became a magnet for domestic population migration, which has led to its intense congestion. Malé is densely populated, accommodating around one-third of the total population (roughly 200,000 individuals) of the Maldives. This situation has led to the necessity of creating another urban area in order to ease overcrowding.

To address the concerns of residents in the Greater Malé Region, the construction of Hulhumalé was undertaken. The main objective was to provide a solution

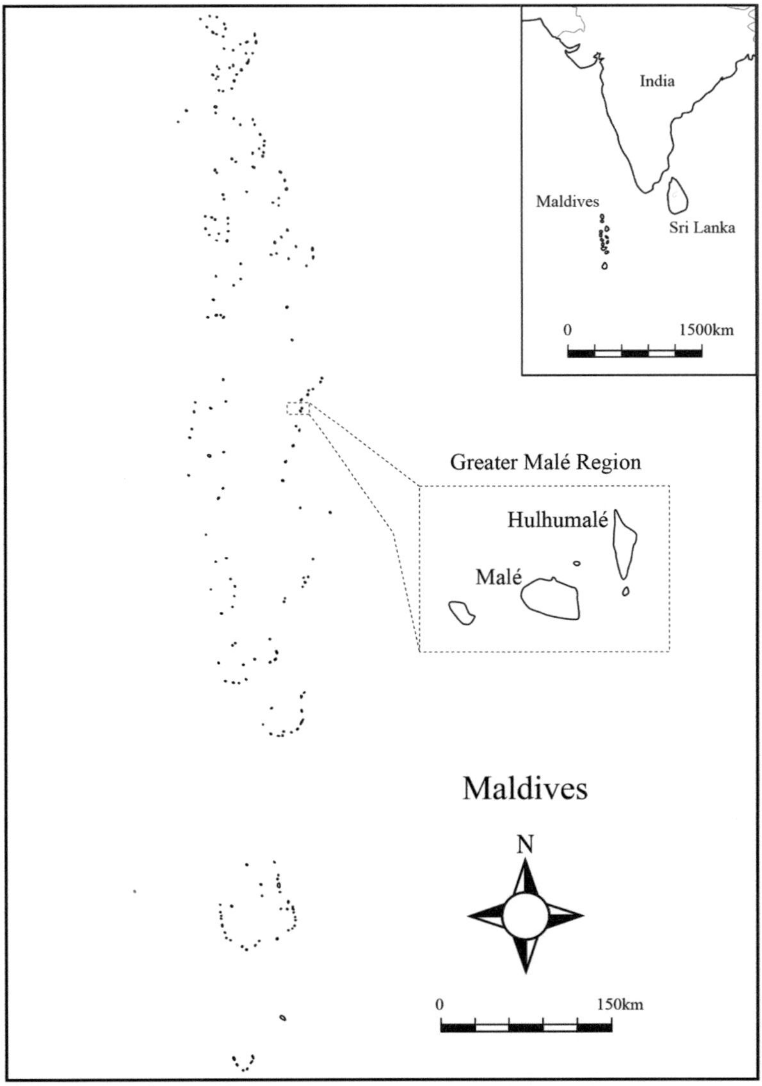

Fig. 2 Location of Maldives and the Greater Malé Region (Malé and Hulhumalé)

by relocating these individuals to Hulhumalé. This initiative aimed not only to mitigate housing-related grievances, but also to incorporate measures for climate change adaptation.

2.3 History of Hulhumalé Development

With criticisms regarding the living conditions of Malé in mind, the government of Maldives has embarked on a unique endeavour: namely, the creation of a new island named Hulhumalé, situated just off its capital. Among the nation's urban centers, Hulhumalé stands as the most developed and thriving urban hub. This transformation is unfolding in two phases: separated by a channel but connected by bridges. The goal of full completion has been set for 2026.

The development of Hulhumalé started on October 16, 1997 (The President's Office Republic of Maldives 2012), when land reclamation work began on the Hulhulé-Farukolhufushi lagoon, which is about 1.3 km off the northwest coast of Malé. The rising sea levels were considered during Hulhumalé's planning stage (BRE n.d.). To address this concern, the island was built 2 m above sea level. The initial reclamation, covering 45% of the total land area (189 ha), was managed by the Ministry of Construction and Public Works. Phase 1's development, including both land reclamation and coastal structure development, was completed by June 2002, with people starting to move in on March 12, 2004 (see Fig. 3).

The subsequent reclamation phase, named Hulhumalé Phase 2, finished in 2015. This phase added an additional 214 ha (about 600 acres) to the island's area, and will reach completion by the end of 2035 (Musmanni 2022, see Fig. 4).

Fig. 3 Central part of Hulhumalé Phase 1 (Photo taken by Akiko Sakamoto on March 19, 2023)

Fig. 4 Residencial area of Hulhumalé Phase 1 (Photo taken by Akiko Sakamoto on March 19, 2023)

2.4 Development of Phase 2 and Future Plan

In 2018, a wave of domestic migration swept towards Hulhumalé; yet the estimated population has not reached the projected figure of 230,000. As of 2021, 35,859 residents called Phase 1 their home (Census 2021). The demand for Phase 2 has been robust, with its construction underway since 2015. Around 10% of Phase 2 has been completed as of early 2023. When combined, the reclamation efforts in Phases 1 and 2 cover a total area of 407 ha. An initiative within Phase 2 is the inception of a mass social housing project, aimed at generating close to 7,000 social housing units that will cater to the housing needs of the population in the Maldives, along with additional housing units for SOE personnel (Musthafa 2018). The blueprint also outlines the creation of commercial spaces.

The grand vision also entails fostering local tourism within the city's boundaries, a strategic move to create an array of job prospects. A cornerstone of this plan is the establishment of an Olympic Level 3 stadium in close proximity to the industrial island Thilafushi, as well as Urban Isle.

Thilafushi is gradually unfolding as an industrial hub, intended to maximize opportunities and forge a synergistic relationship with the international port to be situated at Gulhifalhu. The confluence of these strategic locations aims to amplify the potential of industrial activities housed within the island.

Urban Isle is poised to offer visitors an elevated lifestyle experience: encapsulating a spectrum of accommodation options, premium amenities, and services. This project bears the potential to embody the quintessential 'Maldives' brand, thereby diversifying the nation's tourism industry by introducing an urban dimension to its tourist offerings. This addition aligns with the goal of positioning the Maldives as not only an idyllic beach destination, but also as a thriving urban hotspot.

2.5 *Role of HDC in Developing Hulhumalé*

The Housing Development Corporation (HDC) of Maldives is the leading force behind the expansion, construction, and regulation of Hulhumalé: ensuring the development of essential infrastructure for high-quality living and thriving business ventures. Established in 2001 as a fully state-owned enterprise, HDC originally operated under the presidential office's oversight, focusing on monitoring the growth of Hulhumalé. Subsequently, its name evolved from the Hulhumalé Development Corporation to the Housing Development Corporation in order to encompass a broader range of development projects in the nearby islands of Gulhifalhu and Thilafushi.

HDC is now committed to transforming Hulhumalé into the Maldives' inaugural sustainable city, marking the country's pioneering steps toward intelligent urbanization. The corporation's endeavours encompass a diverse spectrum of projects: including residential, commercial, industrial, institutional, cultural, and hospitality domains.

One advantage of HDC being a company rather than a government agency is its flexibility and efficiency in both decision-making and implementation. As a corporation, HDC can operate with more autonomy and agility, which allows it to respond quickly to market demands, attract private investment, and execute development projects faster and more efficiently. This setup enables HDC to navigate complex business environments, form strategic partnerships, and adapt to changing circumstances more effectivelyy. This stance ultimately accelerates the development of Hulhumalé, as well as the surrounding areas.

HDC's strategic focus revolves around three pivotal dimensions: *Live, Work,* and *Play*. Under the *Live* aspect, HDC is dedicated to both forging sustainable living environments and offering a broad spectrum of housing options. For the *Work* dimension, meticulously designed commercial spaces are complemented by essential services, creating an environment conducive to business operations. Finally, the *Play* facet centers on providing avenues for sports and recreational pursuits. Notably, Hulhumalé emerges as the prime hub for noteworthy events such as marathon races and live music shows, further invigorating its vibrancy.

3 Livelihood in Hulhumalé: Advantages and Disadvantages

3.1 Residents' Assessment of Living in Hulhumalé

The construction of Hulhumalé can only be considered a success if the people who have moved to Hulhumalé feel that life there is more comfortable and satisfying in terms of employment than it was before migrating. To address this, the Maldivian Government is striving to make the lives of those living in Hulhumalé more comfortable while also providing more and better job opportunities.

Preceding Hulhumalé's emergence, the atoll already housed an international airport. Infrastructure like roads were in place, which pre-dates the decision in 1990 to establish a settlement on the island. A notable aspect of Hulhumalé is its connectivity. A well-organized public transportation system, equipped with smart services, is in place; furthermore, a bridge links Hulhumalé to Malé, ensuring easy access to both.

For the purpose of attracting both domestic and foreign migrants, the government of the Maldives launched the idea of making Hulhumalé a Smart City in 2019. Although other SIDS have also embraced this trend, their journey differs due to unique geographical factors, resource availability, such as Mauritius (Tung 2023), socio-economic dynamics, and environmental conditions (Sahib-Kaudeer et al. 2016). The HDC's current focus revolves around smart infrastructure, including an Open Access Network and telecommunication towers.

How the residents rate their livelihood in Hulhumalé is an interesting question. In 2022, HDC and GIF Japan jointly conducted a survey of Hulhumalé residents. The survey revealed the following remarks from 252 respondents (Sasaki et al. 2024).

The environment of Hulhumalé—including its natural beauty, clean air, and water—is generally valued by the locals. In addition to transportation, parks and sports facilities are also highly appreciated. Conversely, the residents expressed the least appreciation for the high cost of living and accommodation, as well as for the dearth of cultural events, well-paying jobs, police protection, and public order. The younger generation is increasingly focused on convenience, including access to government agencies, career and educational opportunities, transportation, and other services. They are also more worried about Hulhumalé's ability to withstand natural disasters. It also appears that the degree of work opportunities is influenced by gender, in that there is a perception among women that they have fewer job opportunities than men. Men value territorial integrity more than women, while women value stunning and diverse natural environments. While inhabitants with lesser incomes feel shut out of services and public facilities, higher income residents take pleasure in Hulhumalé's amenities and facilities (Sasaki et al. 2024).

The residents of Hulhumalé include a mix of migrants from Malé, which is nearby and easily accessible by bridge, and those from more remote areas outside the Malé atoll. How each of these two types of people may assess life in Hulhumalé differently is another interesting question.

Table 1 Difference in opinion among Hulhumalé residents by place of residence before migration

Item assessed by migrants	Place of residence prior to migration			
	Malé		Other than Malé	
	Mean	Median	Mean	Median
High-paying jobs	2.2	2	3.1	3
Non-physical work	2.5	2	3.1	3
Cost of living	2.0	2	1.4	1
Wide choice of entertainment	2.6	2	3.2	3
Rent or house acquisition cost	1.7	1	1.4	1
Clean new home	3.2	3	2.7	3
Quality of primary education	3.1	3	3.4	4
Quality of higher education	2.4	2	3.2	3
Tutoring and learning opportunities	2.7	3	3.3	3
No air and water pollution	3.4	4	3.0	3

As shown in Table 1,[1] for individuals who have relocated to Hulhumalé from outside of the Greater Malé Region, life in Hulhumalé differs significantly from their previous communities, but mostly in a good way. The migrants from areas other than Malé favor the availability of higher paying jobs (with a mean score of 3.1) than those from Malé (with a mean score of 2.2). It means that the need for highly skilled professionals exists only in the Malé area. Not surprisingly, those from areas other than Malé appreciate the quality of higher education and the affluence of learning opportunities in Malé and Hulhumalé (with a mean score of 3.25) as compared to those from Malé (with a mean score of 2.55). In the same context, the availability of a wide range of entertainment is more attractive for those from areas other than Malé (with a mean score of 3.2) vis-a-vis those from Malé (with a mean score of 2.6). Interestingly, those from Malé value the availability of clean new houses more (with a mean score of 3.2) than those from elsewhere (with a mean score of 2.7). Those from Malé value pollution-free air and water in Hulhumalé much more (with a mean score of 3.4) than those from elsewhere (with a mean score of 3.0).

3.2 Issues for Further Consideration and Intervention

Migration to Hulhumalé is not without its concerns. This includes the cost of building the island itself, the price of houses in the area, environmental impacts,

[1] On a 5-point Likert scale, respondents were asked to rate these claims. The options that they were required to select were as follows: "Strongly agree" (5 points), "Agree" (4 points), "Neither agree nor disagree" (3 points), "Disagree" (2 points), and "Strongly disagree" (1 point). The Wilcoxon-Mann–Whitney test revealed a significant difference (p-value < 0.05) between the two groups as shown in table 1.

and possible psychological problems for the migrants. Issues such as these need further consideration and appropriate intervention.

Construction costs for Hulhumalé Phase 2, supported by a US$550 million loan from a Chinese state-owned company contribute to the Maldivian government's housing efforts (Financial Times 2019). The World Bank (2022) highlights those external loans and guarantees for public housing total nearly US$1 billion, equivalent to 19% of GDP in 2020, with 86% owed to Chinese lenders. Despite this, affordable homeownership remains elusive, especially in the Greater Malé Region, leading to a heavy reliance on rental units. However, rental prices often exceed household budgets, resulting in multiple families cohabiting in the same units. As the Maldives transitions from Least Developed Country (LDC) status, citizens share the responsibility of managing costs. Balancing flexibility and accommodation in engineering approaches is crucial for a sustainable urban fabric. The HDC collaborates with partners, securing funds with guarantees due to its State-Owned Enterprise (SOE) status. HDC's income primarily stems from land sales, property sales, and leasing. Buildings are transferred to public entities for management, with payments made to HDC for commercial purposes. Additionally, mobile operators pay rent for using HDC's towers for antennas, contributing to smart city initiatives. Despite significant expenses, HDC is demonstrating its capability and diversification efforts: offering consultation services, tourism and horticulture diversification, maintenance services, and managing two industrial islands.

High housing prices create a significant gap in the region, which is linked to the Maldives's housing subsidy program. Voluntary migration that is facilitated by one's financial means raises concerns about the well-being of those who stay behind. Inclusive social housing initiatives allow everyone to apply, with mechanisms that both ensure eligibility and promote inclusivity. Climate change-influenced migration patterns prompt discussions on holistic solutions beyond technology and financial aid. The ongoing Maldives experiment underscores the urgency of addressing climate change comprehensively, even though technological solutions exist, which require expertise and resources that possibly originate from external sources. The success of such infrastructure projects in one location may be replicated in other countries (e.g., Marshall Islands, Tuvalu, Kiribati) and low-lying areas of non-island states (e.g., Bangladesh). International communities and national governments can contribute through inputs, expertise, and financial support.

Hulhumalé also raises environmental concerns. Known for its vibrant coral reefs vital for biodiversity and tourism, the project's land reclamation and construction activities risk sedimentation and pollution, which can potentially harm marine ecosystems. Studies show minimal reef recovery over a decade, with suggested recovery times exceeding 50 years (Graham et al. 2011). The construction of artificial islands alters natural coastal dynamics, potentially increasing erosion during storms. Hulhumalé's development intensifies pollution risks, impacting water quality and marine life through sediment, pollutants, and construction materials that are released into the ocean. While water quality reduction during dewatering is minimal, population growth and urban development necessitate proper waste management, including hazardous waste separation and disposal.

Despite the fact that SIDS face significant health impacts due to climate change, mental health and well-being aspects within this context are often underreported. Kelman et al. (2021) emphasize the limited attention given to mental health issues in SIDS climate change research. Regardless of the extent of climate change experienced, it is likely to significantly affect mental health and well-being within SIDS populations. Climate change narratives offer an opportunity to address mental health challenges, considering global healthcare system inadequacies and mental health stigma. Stojanov et al. (2017) highlight potential mental health impacts for involuntary domestic migrants in the Maldives who lose their homes and identity. On the other hand, it seems safe to the authors to assume that voluntary migration to Hulhumalé is much less likely to lead to psychological problems such as isolation and loneliness.

While Hulhumalé's open space index compares favorably to Malé, it may not meet other countries' standards. Strategic interconnections enhance its allure, and the city's development prioritizes inclusivity and community well-being.

4 Conclusion

In this chapter, we have described the exciting development of Hulhumalé, a city of hope and a welcome addition to help mitigate Malé's overpopulation issue. With its small land area fragmented over multiple atolls, and with international tourism providing the oxygen to the local economy, the Maldives has no choice but to seriously explore land reclamation. Through the construction of Hulhumalé, the Maldives has demonstrated a proactive response to rising sea levels and to the increasing vulnerability of most of its population to climate-induced hazards. By creating a new urban center with modern amenities and infrastructure, the government aims to alleviate overcrowding in the capital, Malé, while also providing a safe haven for residents at risk of displacement due to sea-level rise. The construction of Hulhumalé made it possible to embed the 'Right to Stay' to its resident population instead of migrating abroad. The phased development of Hulhumalé—spanning over two decades—reflects the careful consideration given to environmental sustainability, community well-being, and economic viability. The incorporation of smart city principles and green infrastructure underscores a commitment to long-term resilience and adaptation.

However, the journey towards sustainable living in Hulhumalé is not without its challenges. Domestic populations will be relocated, state indebtedness will spiral to new heights, and the challenges of starting a new community 'from scratch' will be glaring. The experiences and feedback of residents, as highlighted in recent surveys, provide valuable insights for ongoing improvements and future planning efforts.

The Government of the Maldives, as well as HDC, need to remain vigilant and sensitive to developments in their new creation, with a particular focus on such issues as societal cohesion, mental health, and overall wellbeing. The challenge and experience of the Maldives in building artificial islands as a climate change adaptation measure is neveretheless a source of hope for other small island states, especially atoll countries.

Acknowledgements This chapter is based on the discussions held at a special session convened at Osaka Metropolitan University during the International Conference on Islands and Relations, "Infrastructure development in Maldives to cope with climate change: In-country migration to newly constructed island of Hulhumalé" organized as a side event of "IGU Thematic Conference Osaka 2023" from April 4 to 6, 2023. The authors thank the Global Infrastructure Fund Research Foundation Japan (GIF Japan), Housing Development Corporation Limited (HDC), the International Research Institute of Disaster Science (IRIDeS) at Tohoku University, and the Sasakawa Peace Foundation (SPF).

References

Asian Coalition for Housing Rights (ACHR) (2004) Negotiating the right to stay in the city. Environ Urban 16(1):9–26. https://doi.org/10.1177/095624780401600103

Barnett J, Adger WN (2003) Climate dangers and atoll countries. Clim change 61(3):321–337. https://doi.org/10.1023/B:CLIM.0000004559.08755.88. Accessed 19 Feb 2024

Betzold C (2015) Adapting to climate change in small island developing states. Clim Change 133(3):481–489. https://doi.org/10.1007/s10584-015-1408-0

BRE (n.d.).Hulhumalé, City of Hope. Building Research Establishment (BRE) Group. https://bregroup.com/buzz/hulhumale-the-city-of-hope/ Accessed 19 Feb 2024

Department of State (2003) Compact of free association, 2003. Department of State, United States of America. https://2009-2017.state.gov/documents/organization/173999.pdf Accessed 19 Feb 2024

Djalante R, Lassa S (2019) Governing complexities and its implication on the Sendai framework for disaster risk reduction priority 2 on governance. Prog Disast Sci 2:100010. https://doi.org/10.1016/j.pdisas.2019.100010

Financial Times (2019) The Maldives counts the cost of its debts to China. Financial Times. https://www.ft.com/content/c8da1c8a-2a19-11e9-88a4-c32129756dd8 Accessed 19 Feb 2024

Graham NAJ, Nash KL, Kool JT (2011) Coral reef recovery dynamics in a changing world. Coral Reefs 30:283–294. https://doi.org/10.1007/s00338-010-0717-z

Guadagno L (2016) Human mobility in the Sendai framework for disaster risk reduction. Int J Disast Risk Sci 7:30–40. https://doi.org/10.1007/s13753-016-0077-6

Hioureas C, Camprubí AT (2020) Legal and political considerations on the disappearance of states due to sea level rise. In: Heidar T (ed) New knowledge and changing circumstances in the Law of the Sea. Brill Nijhoff, pp 407–426

IOM (2022) Submission to the mid-term review of the implementation of the Sendai framework for disaster risk reduction 2015–2030. IOM. https://sendaiframework-mtr.undrr.org/publication/international-organization-migration-iom-submission-midterm-review-implementation Accessed 19 Feb 2024

IPCC (2019) Special report on the ocean and cryosphere in a changing climate. https://www.ipcc.ch/srocc/ Accessed 19 Feb 2024

Kelman I, Ayeb-Karlsson S, Rose-Clarke K, Prost A, Ronneberg E, Wheeler N, Watts N (2021) A review of mental health and well-being under climate change in small island developing states (SIDS). Environ Res Lett 16(3):033007. https://pubmed.ncbi.nlm.nih.gov/34149865/ Accessed 19 Feb 2024

Kuyper J, Schroeder H, Linnér BO (2018) The evolution of the UNFCCC. Ann Rev Environ Resourc 43:343–368. https://doi.org/10.1146/annurev-environ-102017-030119 Accessed 19 Feb 2024

Milman O (2022) 'No safe place': Kiribati seeks donors to raise islands from encroaching seas. Guardian https://www.theguardian.com/environment/2022/nov/18/cop27-kiribati-donors-raise-islands-sea-level-rise Accessed 19 Feb 2024

Musmanni GD (2022) Hulhumalé, Maldives: Monumental climate adaptation effort raises hope in a threatened paradise. Global Center on Adaptation. https://gca.org/hulumale-maldives-monumental-climate-adaptation-effort-raises-hope-in-a-threatened-paradise/ Accessed 9 Feb 2024

Musthafa A (2018) Proposed 5,000 social housing units' development project in Hulhumalé' Phase II: Environment impact assessment. http://saruna.mnu.edu.mv/jspui/handle/123456789/3039 Accessed 19 Feb 2024

Nakayama M, Fujikura R, Okuda R, Fujii M, Takashima R, Murakawa T, Sakai E, Iwama H (2022) Alternatives for the Marshall Islands to cope with the anticipated sea level rise by climate change. J Disast Res 17(3):315–326. https://www.jstage.jst.go.jp/article/jdr/17/3/17_315/_article/-char/ja/

RNZ (2019) Marshall Islands considers island build-up for climate change adaption. Radio New Zealand (RNZ). https://www.rnz.co.nz/international/programmes/datelinepacific/audio/2018684823/marshall-islands-considers-island-build-up-for-climate-change-adaption Accessed 19 Feb 2024

Sahib-Kaudeer NG, Jhummun DS, Gobin-Rahimbux B (2016) What is 'smart' for small island developing states? In: 2016 IEEE International conference on emerging technologies and innovative business practices for the transformation of societies. IEEE, pp 13–19

Sakamoto A, Nishiya K, Guo X, Sugimoto A, Nagasaki W, Doi K (2022) Mitigating impacts of climate change induced sea level rise by infrastructure development: Case of the Maldives. J Disaster Res 17(3):327–334. https://doi.org/10.20965/jdr.2022.p0327

Sasaki D, Sakamoto A, Laila A, Aslam A, Feng S, Kaku T, Sasaki T, Shinomura N, Nakayama M (2024) Facilitating the smooth migration on inhabitants of atoll countries to artificial islands: case of the Maldives. Sustainability 16(11):4582. https://doi.org/10.3390/su16114582

Schipper ELF (2009) Meeting at the crossroads?: Exploring the linkages between climate change adaptation and disaster risk reduction. Clim Dev 1(1):16–30. https://www.sei.org/publications/meeting-crossroads-exploring-linkages-climate-change-adaptation-disaster-risk-reduction/

Stojanov R, Kelman I, Procházka D, Němec D, Duží B (2017) Climate change and migration in Maldives. Georgetown J Int Aff 3 https://www.stojanov.org/file/gjif-maldives.pdf Accessed 19 Feb 2024

The President's Office Republic of Maldives (2012) Cabinet discusses plans for Hulhumalé development project. Press Release, Ref: 2012–251. https://presidency.gov.mv/Press/Article/7128 Accessed 19 Feb 2024

Tung OJ (2023) Smart city regulation and environmental sustainability in the context of land use planning in Mauritius: a critical review. Potchefstroom Electron Law J 26(1):1–39

UNISDR (2015) Sendai framework for disaster risk reduction 2015–2030. UNISDR. https://www.undrr.org/publication/sendai-framework-disaster-risk-reduction-2015-2030 Accessed 19 Feb 2024

White House (2021) Report on the impact of climate change on migration, October 2021. https://www.whitehouse.gov/wp-content/uploads/2021/10/Report-on-the-Impact-of-Climate-Change-on-Migration.pdf Accessed 19 Feb 2024

Wille B (2023) Atoll engineering in the Maldives: shifting priorities in crafting archipelagic landscapes. Int Rev Environ Hist 9(1):75–99

World Bank (2022) Maldives development update: towards resilient and affordable housing. World Bank. https://thedocs.worldbank.org/en/doc/49141824db6f3a3b812bad33d5779541-0310062022/original/MDU-Oct-22-Final.pdf Accessed 19 Feb 2024

Yamamoto L, Esteban M (2010) Vanishing island states and sovereignty. Ocean Coast Manag 53(1):1–9. https://waseda.elsevierpure.com/en/publications/vanishing-island-states-and-sovereignty Accessed 19 Feb 2024

Baseco Compound: The River Island of Contestation, Despair, and Hope in the Philippines

Joefe B. Santarita and Godfrey Baldacchino

Abstract Baseco Compound, an artificial island in Manila, the Philippines, was built as the dockyard of the National Shipyard and Steel Corporation in the 1960s. It was a sizeable 30 (and now 48) hectare waterfront, and borders the Pasig River and Manila Bay after reclamation. From being the site of government facilities that repaired ships and retooled heavy and light machinery, the island was reclaimed through the years to become the Bataan Shipyard and Engineering Company's (Baseco) dockyard. Owned by the kin of former First Lady Imelda Romualdez Marcos, Baseco was subsequently sequestered and nationalized by Corazon Aquino's administration in the 1980s. The area was transformed into the resettlement site of many families that had been displaced by Ferdinand Marcos' clearing operations in Metro Manila to give way to the construction of government projects. The newly arrived families erected their makeshift houses on a low-lying and unstable substrate that is a combination of river silt, sand, rubble from old buildings, dredged mud, garbage, and small white mollusks. During high tide, these houses would be flooded with seawater. Recently, the site has become the home of 60,000 urban poor and a pockmark of Metro Manila. This chapter looks at the transformation of the island as defined by anthropogenic forces (such as river flooding, high tide, land reclamation, garbage and many more), with information and insights drawn mainly from archival and library materials, contextual literature to artificial and river islands contributed by the second author, as well as from interviews undertaken by the lead author with a convenience sample of respondents during March 2023 and February 2024. It also examines the island's role as a source of political, economic and environmental contestations, as well as of both despair and hope to inhabitants and local government units.

Keywords Artificial island · Baseco · Metro Manila · Philippines · River island

J. B. Santarita (✉)
Asian Center, University of the Philippines, Diliman, The Philippines
e-mail: jbsantarita2@up.edu.ph

G. Baldacchino
Department of Sociology, University of Malta, Msida, Malta

T. Yamazaki and G. Baldacchino (eds.), *Relational Geographies of Islands in the Indo-Pacific*, International Perspectives in Geography 23,
https://doi.org/10.1007/978-981-96-7155-7_3

1 Introduction

Artificial islands are manifestations of the confluence of natural and human environments, blending elements of geography, ecology and human ingenuity; and—in the case of Baseco Compound—sheer necessity. The construction of artificial islands, led by human-driven reclamation technology rather than formed through natural processes, can be traced back to the period of ancient Egyptians. This common civil engineering technique gained global attention when China started constructing its first artificial island in August 1995 (Sreekumar and Hassan 2020). Since then, China has continued building more such islands, even in contested waters. Today, artificial islands can support entire cities, such as the 970 km^2 Flevopolder in The Netherlands, currently the world's largest island formed by reclaimed land. Inhabited or uninhabited, temporary or permanent, all artificial islands bear the imprint of human activity, and are dismissive of 'nature' and culture (including technology) as distinct ontological zones. They include the high profile infrastructures of Kansai International Airport in Japan; and the Palm Islands of Dubai, United Arab Emirates (Dodds and della Dora 2018: 394–396; Jackson and della Dora 2011).

In modern times, the building of artificial islands is triggered by several compelling factors such as the need for residential, industrial, commercial, and even strategic purposes. Erected in various shapes and sizes, these islands are usually made using highly sophisticated machinery, technology, and engineering skills. Moreover, these islands are constructed through one or more methods: land reclamation; constructing or extending over existing islands, rocks or coral reefs; or linking islets by filling the in-between areas using different construction materials (Sreekumar and Hassan 2020). The development of Bataan Shipping and Engineering Company (Baseco) Compound was a product of the combination of reclamation, extension over an already existing rock base, linking to the shallow areas of Manila Bay, and even integrating accumulated garbage along the same bay.

Aside from being an artificial island, Baseco Compound in Manila, the capital city of the Philippines, falls into the category of a river island. The river island is an exposed landmass which through time will assume various shapes because of the changes in the currents and intensity and direction of the flow of the river. This alteration in the shape of the landmass has a significant influence on the river bed, water channel, ports, embankments, as well as ecosystems and other Anthropocene activities on the island (Shi et al. 2017).

Baseco Compound, however, is distinct in its own right. Compared to other river islands in the world, particularly in Southeast Asia, it is mostly a human-made island located at the confluence of a river and the sea. Baseco Compound was also called Engineer Island because it served as the site of government facilities that repaired ships and retooled many heavy and light machines. This artificial island, which formerly covered just about 30 hectares, is now known as Baseco. Through the years, the island has expanded, with two additional stone breakwaters. One of these breakwaters borders Manila Bay's South Harbor while the other straddles the Pasig River. Prior to the 1980s, these breakwaters were two separate islands, called Big

Island and Small Island. After the 1986 People Power Revolution, Baseco's area has expanded with a series of reclamation works initiated either privately or jointly with the government, residents themselves and/or with the help of non-government organizations (NGOs). Interestingly, the area has its own beach which is visited by several Manileños despite the danger of high presence of coliform in its surrounding waters. The area is also prone to flooding, especially during the typhoon season and strong monsoon rains. The island also bears the infamous reputation of being a 'haven of criminals and other shady characters.' Aside from that, it is also an open secret in the country that the island has once been the source of human kidneys as commodities (Aravila 2001).

Because of such distinct characteristics, the island is the focus of this chapter. Most of the material has been drawn from archival research, specifically from historical documents available online, as well as from ethnographic field research carried out by the first author (Santarita) during his courtesy call to the barangay officials and follow-up sessions and interviews with residents in March 2023 and again in February 2024. Santarita took photographs of several establishments in the Baseco compound in early 2024. Additional suitable imagery, contextual literature, and general reviewing and editing have been provided by the second author (Baldacchino). In particular, this chapter looks at the island's transformation as defined by the amalgam of anthropogenic and natural forces: the river's flooding, high tide, reclamation, garbage, government projects, and more. It also examines the island's role as a source of political, economic and environmental contestations, as well as of despair and hope to its inhabitants and local government units.

2 Baseco, the Manila River Island

2.1 Geography

What was formerly the Baseco compound can be seen from the air today as a 'v' formation and is composed of Engineer Island and two stone breakwaters nearby (see Fig. 3.1). As mentioned earlier, one of these structures borders the Pasig River while the other faces Manila Bay. Schools, a health center, a hall for the barangay (or barrio, an administrative district), a police sub-station, hospitals and shipping facilities are located at the base of the 'v' area where Engineer Island was originally located (see Fig. 3.2). The area is also divided according to social status. The cemented houses of better-off families are constructed on the island, while the dwellings of much poorer residents are built beside the two breakwaters along the bay. These families reside in houses built on stilts over shallow water, packed close together like driftwood piled up on the beach. The only access deep into the blocks is along the huge stones that form the breakwaters that surround a lagoon (Murphy 2012). This body of water, however, was gradually reclaimed after 2001 and completely covered within the span of four years (Valenzuela et al. 2020). The reclamation process was also hastened by the accumulation of plastic waste freely carried downriver. These wastes formed a bouncy mat on the water which is thick and hard enough to establish a hut on top (Choi 2021). In recent years, Baseco has extended more than 0.6 km^2 from its original size as a flood protection barrier (Cutini et al. 2020).

The inhabited area along the South Harbor breakwater is known as Big Island. This contains Blocks 1 to 14 and covers almost a kilometer along the western front. The piles of houses near the breakwater along the Pasig River are called Small Island. This area covers Blocks 15 to 17A, and is not as long or as heavily populated as the other. Joining the two areas along the breakwaters to the west of the river island is a beach of white sand. This beach is called "La Playa" where the local government unit of Baseco has managed to discourage people from constructing more houses. One reason is to protect the mangroves located here, as well as because of the area's high vulnerability to disaster, given that most houses are built out of light materials and are destroyed easily by typhoon or just heavy rain (Murphy 2012). The local government also seriously discourages people from swimming in the adjacent water of the beach because of the presence of intolerable level of coliform bacteria; and yet, some locals still find the water irresistible.

Several initiatives have been taken in recent years to protect the area from storm surges, flooding, and environmental degradation. Aside from regular coastal clean-ups, the government has constructed the boardwalk to improve the physical environment as well as to monitor the progress of mangrove plantations inside the nearby lagoon. The 348-m boardwalk was also constructed to encourage eco-tourism, coming with the establishment of a view deck that has a carrying capacity of about 30 individuals (Mayuga 2020). As of February 2024, certain areas are cordoned off by the authorities to ensure that the growth of newly planted or recovering mangrove saplings is not disturbed by unwarranted entities.

2.2 *Contestation*

At some 6.7 m above mean sea level, the island has been an arena of contention and contestation due to its geographical and demographic nature, its location and the sheer number of people that now call it home. Such characteristics cumulatively affect the political and socio-economic situations of the inhabitants and of the sustainability of the island.

In terms of political geography, Baseco Compound secured a political identity in the 1980s as its own administrative unit, known as Barangay 649 (PhilAtlas 2024). Given that the island is at the confluence of the swelling Pasig River and Manila Bay proper, its inhabitants live as permanent hostages to the wrath of meteorological conditions, particularly during typhoons and heavy rains; and the rise of sea level in Manila Bay will, slowly but surely, exacerbate their predicament.

In terms of demography, the island is the home of 64,750 inhabitants according to the 2020 Census, which is 3.5 percent of the total registered residents of Manila (Philippines Statistical Authority 2020). Barangay 649 is also the most populous

Fig. 3.1 The location of Baseco in Metro Manila (*Source* Takashi Yamazaki based on OpenStreetMap)

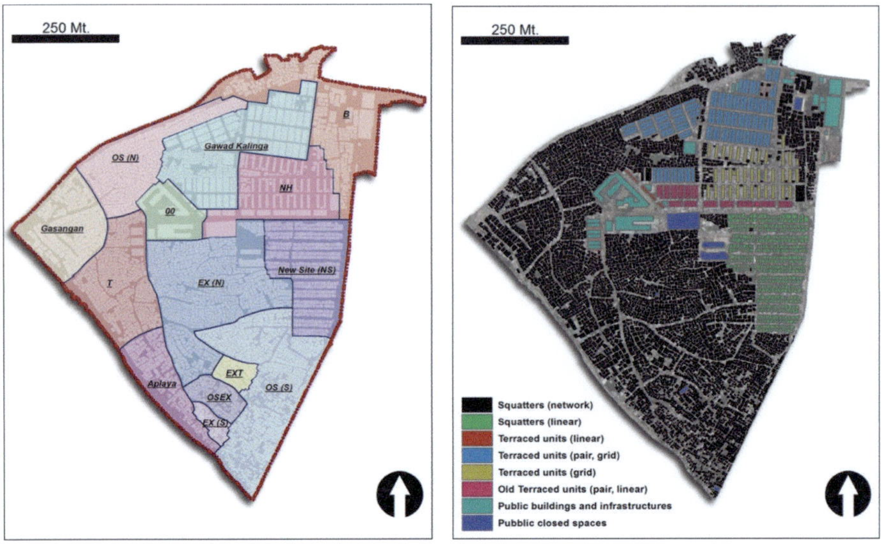

Fig. 3.2 Urban morphology in Baseco: urban districts (left); building types and related morphology (right) (*Source* Cutini et al. 2020. Reproduced with the kind permission of the copyright owners)

administrative unit in the city which is a staggering 32 times denser than the average barangay inhabitants of about 2,000 souls in Manila (PhilAtlas 2024). The population density of Baseco is estimated at 1,295 people per hectare. The denser location of dwellings creates tensions between neighbors over water and electricity supplies, as well as waste disposal. It is a poorly reclaimed surface of about 48 hectares in size and with almost 10,000 housing structures surrounded by rubbish all over (UNDESA 2019). Since the reclamation was done 'in various stages through changing political landscapes, and much of the fill area was improperly mixed with garbage and sand, leaving the base unstable and prone to liquefaction during seismic events' (Smith 2019). The unstable foundation also leads to occasional dump slides and a high likelihood of the presence of combustible methyl gas, generated by dumps, and which could cause fire (Soriano 2003). Because of its poor and ramshackle base and infrastructure, the area is also highly vulnerable to typhoon damage, flooding, fire, earthquake, and other natural and human-made disasters. The compound is the home of almost 13,000 families, and about 45 percent of residents are children. Residents consist mostly of informal settlers with low income, insecure jobs, and limited or no access to basic services such as proper shelter, health care, sanitation, and primary education (UNDESA 2019).

The primary sources of livelihood of 65 percent of the Baseco residents are informal, ranging from fishing to vending, scavenging, tricycle and pedicab driving, and stevedoring. The other sources of employment are construction, truck driving, security, overseas Filipino workers (OFW), and government jobs. Many of them have provided their services and traded their wares within the island, while some are employed as government personnel and, therefore, would need to travel for at least an hour to and from nearby cities in Metro Manila for many days per week. A good number of residents are also working abroad as OFWs. Despite the latter's economic mobility, many have decided to stay in the area because of their familiarity with the neighborhood and logistical advantages with relatives. 'For those who have a more formal occupation, these residents tend to earn more but remain in Baseco due to a lack of better housing options and cheaper rent' (de Guzman 2022: 43).

For many years, Engineer Island had been a ship loading and unloading site as well as a ship repair yard. Once managed by the National Steel and Shipbuilding Company, the site was obtained in 1964 by the Romualdez family, who named it as Baseco. As a private firm, the area around the shipyard in the 1960s and 1970s was populated by workers and their families. Moreover, the island occasionally became a place of respite for fishers from Bataan before they proceeded to the fishing grounds in Cavite.

When Baseco was declared a barangay in the 1980s, there were only 15–20 families residing in the area. However, when President Corazon Aquino came to power in 1986, hundreds of families moved in. It should be noted that the area in the late 1970s was formerly cleared of urban poor settlers by then President Ferdinand Marcos. The clearing was instigated by the need to create space for facilities required for Manila's new international seaport terminal. After 1987, thousands of families reoccupied the area (Murphy 2012).

The arrival and settlement of many new residents in Baseco bloated its population density. There were families who had just come from the outlying provinces and joined their friends or relatives in Baseco. A large number of residents were believed to be either relocated to the compound from other communities that had been cleared in Metro Manila or were brought in by professional squatter syndicates. This re-occupation of land has consequently expanded the reclamation activities in the area. The newly arrived families constructed their makeshifts on low-lying surfaces made of rubble from old buildings dumped by the Department of Public Works and Highways (DPWH), mud dredged from Manila Bay, river silt, garbage, and *gasang-gasang* (small white shelled mollusks). Most of the houses at that time were built on stilts over shallow water, which made them vulnerable during high tide when water levels reach the houses' lower floors. The area is dynamic: over time, garbage, construction debris, and sediment continue to pile up as its substrate, allowing the formation of additional terrain which becomes the site of abodes for new settlers (Murphy 2012).

Migration to the island increased in 2002 when President Gloria Macapagal-Arroyo promised the land to "current occupants" in a bid to secure their political support. Two years later, Arroyo's government resorted to landfill engineering in order to reclaim the lagoon surrounded by adjacent breakwaters (Feng and Hwang 2022).

Aside from the issue of ownership, there are other tensions that grip the island's inhabitants, such as between the older and more recent migrants, as well as some tensions between Christian and Muslim refugees from Mindanao, Southern Philippines, in the early phase of migration. By 2007, "Baseco had nearly 6,000 Muslim residents. They were able to settle there from the provinces in Mindanao or other parts of Metro Manila mostly through family connections, attracted by job opportunities in the shipyard" (Watanabe 2007).

2.3 *Despair*

Aside from these areas of contestation, several conditions present in the river island also result in despair. Baseco at first sight "presents all the typical characteristics of any informal community such as congestion, poor hygiene conditions, misaligned street spaces and lack of urbanity" (Cutini et al. 2020).

As mentioned, the abject living conditions of the residents are noticeable. A good number of inhabitants in the islands are migrants from provinces whose primary motivation is to look for better prospects in Manila. Almost half of Baseco's current population is composed of Muslim migrants who fled the war-torn municipalities in Mindanao in the 1990s. Equipped with hopes and dreams, these people came with high spirits to the metropolitan area, but they lacked marketable skills, qualifications and education. They end up jobless, vulnerable and in precarious conditions. In their first years in the compound, many Muslim migrants also encountered exclusion and discrimination, which pushed these to engage in illicit activities (International Alert 2019).

Meanwhile, most males seek to earn some income by driving tricycles, pedicabs, jeepneys, buses, trucks, and taxis; or by serving as boat operators, security guards, or hawkers in nearby Divisoria market. Women, on the other hand, usually become domestic helpers, engage in sidewalk vending, hawking, or running sari-sari stores and mini-enterprises. Such poor conditions have even led desperate male residents to sell their organs in exchange for money. For a certain period, Baseco was once notorious as a hotbed for the kidney black market. This illegal and lucrative trade was exposed and sensationalized on national television in early 2000 when several Baseco residents sold their organs to local and foreign buyers. Local officials estimated that around 3,000 Baseco residents had sold their kidneys. The asking price for a kidney based on online advertisements can go up to half a million pesos (around US$8,875), depending on the parties' negotiation skills (Promchertchoo 2019).

Another cause of despair concerns the safety conditions of the residents. The limited access to clean potable water increases exposure to untreated human waste and consequently affects the residents' hygiene and health situations. Such poor sanitation practices have triggered the increasing transmission of diseases (e.g., diarrhea, cholera, worm infections) and exacerbate stunting and malnutrition. The poor sanitation practices include the direct disposal of human waste in the ocean, where houses and makeshifts situated along the coastline have toilets located directly above the ocean. Beaches used to be areas of open defecation (Choi 2021). Because of this, the authorities turned away guests to Baseco Beach due to the high levels of fecal coliform in the compound and in nearby areas.

More misery on the island results from the frequent occurrence of fire incidents in the area. Fire incidents have happened at least five times a year: caused by a cigarette, an open gas range or faulty electrical wiring, and which swiftly develops into a fire of uncontainable proportions. These uncontrollable fire cases are caused by the cheap and flammable nature of materials used to construct housing such as cardboard, tin, and corrugated steel, among others (Choi 2021). There were years when the blaze left 4,000 to 25,000 homeless due to undetermined reasons, particularly in 2002 and 2004 (Cerojano 2010). In the last month of 2023, another fire incident took place in the residential block of the Gasangan area (Delizo 2023). Moreover, the devastation wrought by fire incidents is exacerbated by the absence of local water sources and the non-availability of fire extinguishers to control the spread of fire while it is still manageable (Melican 2015).

Furthermore, as if these maladies were not enough, another source of despair in the island is the threat of storm surges, or huge waves inundating the land during storms. Many areas in Baseco become flooded even if there is no typhoon, such as during strong monsoon rains, when residents unfortunately experience knee-deep floods. If there are storm surges, floods can reach waist-deep levels (Melican 2015).

3 River Island of Hope

3.1 Presidential Proclamation and Government Initiatives

However, this is just one side of this poignant human story. Aside from the contestation and despair, the island also promises hope to its inhabitants. With the support of the government offices and the assistance of various private entities and NGOs, the residents on the river island have glimpsed a proverbial ray of hope at the end of the tunnel.

The first sign of hope was the presidential proclamation of the area as the site for social housing in 2002. Although this was seen as a political gave-away of then President Arroyo to appease poor urban residents who were supporters of the deposed President Joseph Ejercito Estrada, the proclamation on February 12, 2002, was good enough to give the inhabitants the 'security' of land tenure. Several factors prompted such a proclamation. First is the security of land tenure that has been emphasized by the 2001 study commissioned by the Asian Development Bank (ADB), and the Pasig River Rehabilitation Commission (PRRC) as a high- priority concern for urban renewal. This becomes the prerequisite condition for ADB's urban renewal work and financing. Second is a form of compromise by giving the inhabitants a piece of land to soften their loyalties to former President Estrada and win them to support President Arroyo. Third is the organization of *Kabalikat sa Pagpapaunlad ng Baseco* (Kabalikat), a people organization, in Baseco in March 2001 which seriously considered the land proclamation as one of its advocacies. The civil society organization has lobbied the president and consequently got the nod of Manila City Mayor Lito Atienza. Fourth is the concurrence of government entities to support the proclamation. The Philippine Ports Authority (PPA) managed the area on behalf of the Philippine Estates Authority (PEA) and the deliberate go-signal from Mayor Atienza has hastened the land grant to the residents (Murphy 2012).

Aside from the proclamation, several initiatives were also implemented by various government offices in the area. These agencies assisted the barangay council and the residents to sustain their living conditions and upgrade their working and living environment. These include the Metropolitan Manila Development Authority (MMDA), the Department of Environment and Natural Resources (DENR), and the City Government of Manila, and others.

Foremost is the DENR. Its National Capital Region (NCR) office serves as the lead agency in rehabilitating Baseco. At the outset, the Manila Bay Rehabilitation in collaboration with the MMDA aims to clean up the Baseco lagoon and its environs and to conduct tree planting. Since 2019, the regional office has already planted thousands of seedlings comprising various species of mangrove and beach-type trees. The survival rate is reported by the national agency to be 87 percent (Department of Environment and Natural Resources (DENR) 2020). The DENR-NCR and MMDA along with other NGOs have started and continue to support the mangrove plantation in Baseco. Mangrove reforestation as a strategy is designed to improve water quality in the area as well as putting an added layer of protection against geological and meteorological hazards, especially storm surges and tsunamis (Mayuga 2020). The confluence of humans and nature is best demonstrated with the implementation of dredging projects at the mouth of the lagoon in 2021 to restore the inflow and outflow of water in the area. It should be noted that mangroves need to be exposed to tidal waters. Due to layers of trash and other debris, the small channel to the lagoon prevents water circulation. Hence, dredging is necessary (Hrvacevic 2022).

Aside from habitat restoration through tree planting, the DENR-NCR has also initiated the construction of a boardwalk in the Baseco area. Moreover, several Estero Rangers under the employ of DENR-NCR are deployed in the area to conduct daily clean-up activities along the 'la playa' section of the Baseco beach and surrounding area (Village Connect 2021).As a result, the Baseco beach has been cleared of floating garbage. Although the coliform level has decreased significantly due to several rehabilitation initiatives, the waters remain unsafe in Manila Bay (Ong 2023). Baseco inhabitants, especially children, are discouraged from swimming off the beach (Village Connect 2021; also Edmundo Cayanan, Interview).

The local government unit of Manila City also supported the residents by providing 229 townhouse-style units to its indigent owners through a project called 'Basecommunity' (Fig. 3.3). Each of the 42 square meter units contains two bedrooms, a living room, a kitchen, a laundry area, and a toilet (Mendoza 2021). At the same time, the local government under the administration of Mayor Francisco Domagoso constructed the President Corazon Aquino General Hospital to assist the health needs of the residents in the barangay (Diana Espinosa, chairwoman of Barangay 649, interview). This three-story facility with a 50-bed capacity is designed with separate male and female, isolation, and pediatric wards as well as a fully operational emergency room, a centralized oxygen supply line, and digital X-ray equipment (Moaje 2021;

Fig. 3.3 Basecommunity (Photo by Joefe Santarita, January 2024)

see Fig. 3.4). To augment the delivery of basic health services in the island, various offices also offered primary health care services to the residents. One of these is the government initiative led by First Lady Liza Marcos that made Baseco as the launching pad of her Laboratoryo, Konsulta at Gamot (LAB for ALL) program. This program basically offers free laboratory testing services for a variety of diseases, medical check-ups, family planning services, and distribution of free medicines to around 1,500 beneficiaries in the area (De Leon 2023).

3.2 *Supporting Organizations*

Apart from government support, several NGOs have been assisting the island's inhabitants. These include the Urban Poor Associates (UPA), Caritas Manila, Habitat Philippines, Rotary Club of Manila, Ummahat Baseco (Muslim Women Organization), Bless the Children Foundation, Grundfos Foundation, and World Vision, among others. These organizations have provided the residents of the island several radiant of hope, ranging from housing to job opportunities and income generation activities, facilities for material resources, maintenance of peace and order, provision of potable water, and educational support to young children. Hope is also seen through the long term doable and sustainable activities such as work-related training, harvesting homegrown vegetables as well as planting mangrove saplings in the area.

In terms of housing, the *Habitat Philippines NGO* has provided safe, decent, and affordable homes to Baseco residents. The Bagong Buhay Village in the area has its

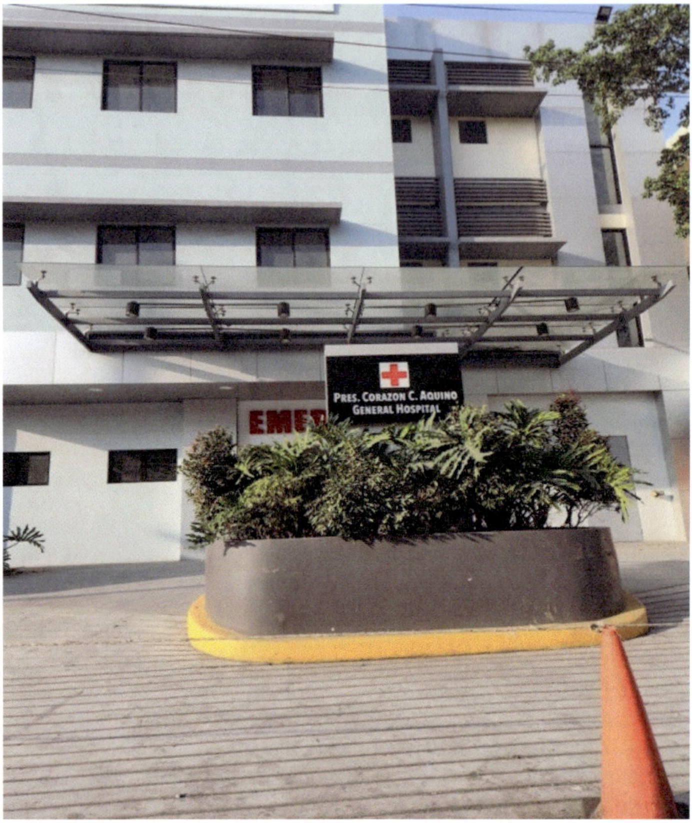

Fig. 3.4 President Corazon C. Aquino General Hospital (Photo by Joefe Santarita, January 2024)

own commercial center which provides livelihood opportunities, a community center for strengthening ties, and a worship center for the residents' spiritual needs. Due to the soft soil at Baseco, the Habitat imported steel-frame technology from New Zealand. The light-weight steel frames as well as fire and termite-resistant cement walls were used in building 24–30 square meter housing on the island (Habitat for Humanity 2007). This NGO has already built thousands of housing units in Baseco to date.

With regards to grassroots empowerment, the *UPA* works with the residents to search for alternative ways to address their problems. This NGO has organized Kabalikat in 2001 which advocated for the security of tenure in the compound. This organization is composed of empowered individuals in Baseco who were trained to negotiate with the government and other offices to solve their woes and offer tangible solutions. Kabalikat now runs numerous programs that serve as the binding factors for its leaders and partner families. These include training in livelihood management, environment protection and mitigation prevention, a financial savings scheme, health, security of tenure, and provision of small infrastructures. A number of unemployed adults are now being helped by the Kabalikat program through bag and handicraft making from scrap materials and water lilies found in the Pasig River. Habi bags are now being marketed in Kultura boutiques in several SM supermalls in the country. The group also helps poor children who are slow learners in their classes to fully understand their lessons in Math, English, and Science through tutorials. Kabalikat also implemented the mangrove rehabilitation project along the seashore of Gasangan in Baseco (Urban Poor Associates 2022; also, Norberto Papa and Rolie Roldan, Interview). It also partnered with Pilmico Food Corporation to support Kabalikat's weekly feeding of over 200 grade school students, free medicine, and medical assistance on the island (Pilmico 2018).

Hope is also seen in the concerted actions of the residents, particularly those who were formerly marginalized in the area, as demonstrated by the *Ummahat Baseco*. This is a group of Muslim women who have been advocating since 2007 for the provision of basic services in Baseco. They formed the Ummahat Manila Enterprise (UMME) which seeks to raise funds for the operation of Mahad Baseco Madrasa (an Islamic school) and facilitate remunerated Islamic teaching. These teachers educate young Islamic adherents in the compound to be resilient in the face of the enticement of illegal activities, extremist ideologies, and recruitment to criminal gangs (International Alert 2019). On top of this, the group has also implemented a 'rice for peace' social enterprise that aims to provide additional sources of income for women and sustain the community-run madrasa.

Moreover, hope is manifested in the environmental and sustainable programs initiated by religious and private institutions. Foremost is *Caritas Manila*. The social services and development ministry of the Catholic church in the Philippines has built the Material Recovery Facility (MRF) in collaboration with Coca-Cola Philippines in the compound. MRF was erected to generate income for the residents, especially women, by collecting and selecting wastes that can be reused and sold. By doing so, this initiative will also reduce the volume of garbage dumped on the island. MRF is benefitting around 600 families (Prado 2022).

Another player is the *Rotary Club of Manila,* which supported a garden project for struggling families living in the compound. This garden is designed to serve as a beacon of hope to 20 volunteer families in Baseco. In January 2021, the volunteers were able to grow 426 small to medium-sized items of 12 different kinds of vegetables (Rotary 2020).

Aside from waste management and garden projects, the residents also benefitted from the generous support of *Grundfos Foundation*. This Danish foundation has installed US$130,000 worth of solar- powered water filtration units with the help of the Asia Society for Social Improvement and Sustainable Transformation (ASSIST) in Baseco. These water treatment systems have provided 96,000 L per day of clean water for drinking and hand washing to some 4,000 Baseco residents (Tiangco 2020).

The island residents have also received generous donations from the business sector to protect the environment. The ABS-CBN Lingkod Kapamilya, in cooperation with the Pasig River Rehabilitation Commission (PRRC), for instance, set up a mangrove farm in a slum area and has consequently turned this into a tourist spot (Clapano 2015). Because of such initiatives, along with clean-up drives, fishers are also now surprised to start catching again assorted kinds of fish in Baseco's waters. Inhabitants are now also catching shrimps using fine nets, which they either sell or cook for themselves as a food item (Ancheta et al. 2019; also Ian Pido, Interview).

The local government is also partnering with business enterprises in the clean-up drives. One of them is the Alaska Milk Corporation, which 'brought together volunteers, local residents, and various stakeholders who took part in cleaning up the coastline, collecting and documenting trash, and raising awareness about the detrimental effects of pollution on marine life' (Alaska Milk Corporation (AMC) 2023).

The residents also found hope from the training and assistance provided by several organizations to generate jobs and income. The *Bless the Children Foundation*, for instance, has organized the Smokey Tours to help residents generate income from the resources available on the island. The tour encourages interactions by allowing guests to be invited inside the residents' homes and engage in conversation with them. Some guests would return home with a new perspective on life; others would stay and become volunteers for the tours. At the same time, the foundation trains selected locals as tour guides, which enables them to earn something without leaving the island and their families. These locals make the best guides since they have insider knowledge of how the community really works. Furthermore, women were educated to make slippers, bags, and accessories from the stalks of water lilies that are abundant in and around the island. These products are then sold in the souvenir shops of the foundation (Filart 2021). The proceeds generated from the tours and donations greatly contribute to financing various projects, such as the opening of a health center in 2017, health programs (providing vitamins, midwife assistance, and proper guidance on nutrition and breastfeeding) in 2018 for pregnant ladies, and a talent center in 2019 that will assist residents to make their resume, and application letter as well as prepare for the interview of a potential job offer (Bless the Children Foundation 2024).

Parents and younger residents also see hope in the continuous developmental programs of *World Vision* on the island. This global humanitarian organization has provided aid to displaced families due to fire, has educated people on how to turn their wastes into wages by rewarding recycled grocery packaging and those who collected Polyethylene terephthalate (PET) type of plastics in the area, and has sponsored the entrepreneurial and employable skills training to out of school youth (World Vision 2019, 2020). It also partnered with Xiaomi (a technology company) to donate smart devices (air purifiers, smart pads and televisions) to Baseco High School to assist students and teachers in their Grade 12 ICT curriculum (Samaniego 2023).

By and large, the above-mentioned institutions have offered the residents considerable opportunities and sober optimism in various means and modalities. Residents are being taught and supported to stand on their own: something akin to the Chinese proverb "*teach a man [sic] to fish and you feed him for a life time*". In a similar vein, the NGOs and government offices have gone beyond dole handouts, donations, and assistance, via training and other self-empowering programs. These activities seek to protect the environment and consequently make it sustainable through mangrove plantations. By doing so, these programs cumulatively allow the residents to harvest marine resources in the area in the long run and shield them at the same time from flooding and storm surges during typhoons. Hope is also seen through the educational support provided by several organizations to the large young population of the compound. Many Filipinos widely see education as a means to escape from poverty. Hence, individual and institutional educational investments are clearly observed and appreciated in the area, as much as in the Philippines as a whole.

4 Conclusion

An island that arises at the edge of one of the world's largest metropolitan areas is bound to reflect an amalgam of extreme dynamics: the sheer mass of humanity; the sheer acts of desperation; mostly informal; some criminal—that residents adopt in order to survive; and the multiple attempts by the state and its affiliates, as well as NGOs, to provide some solace to inhabitants, or to 'manage' and administer the island, in spite of its challenging and overwhelming circumstances. Here is an island that defies clear definition, but is obviously too weighty to ignore. Baseco island is urban, river and artificial rolled in one; an amalgam of identities that corresponds to the mish-mash of its substrate layers as well as the diversity of its human occupants.

Being the most densely populated area in the country's largest and most densely populated city, Baseco island has suffered the worst from environmental degradation and climate change compared to nearby districts in the mainland of Manila.

Because of its notoriety, and perhaps its vote-rich importance to politicians, the artificial island has gained the attention of state and non-state actors as the targeted recipient over time of various developmental programs, including environmental protection, sustainability, circular economy, and social justice. It is important, however, that such actions and initiatives should be sustainable in the long run. The island needs the help of all stakeholders. Inhabitants can do their part by considering waste management seriously and actively collaborating with various local government units, national departments, private industry and civil society. Only then can Baseco, as a river island, become a place of hope for those who deal with it or dwell on it.

References

Alaska Milk Corporation (AMC) (2023) Alaska Milk partners with Manila City for International Coastal Cleanup Celebration 2023. https://www.alaskamilk.com/purpose/newsroom/alaska-milk-partners-with-manila-city-for-international-coastal-cleanup-celebration-2023/ Accessed 24 Sep 2024

Ancheta A, Membrebe OZ, Valeroso JC, Jomarie A, Santos G (2019) Framing disaster among the locals in urban coastal communities along Manila Bay. J Nat Stud 18(1):38–51. https://www.journalofnaturestudies.org/files/JNS18-1/38-51_Ancheta_Framing%20Disaster%20Locals.pdf Accessed 24 Sep 2024

Aravila J (2001) Barangay Baseco: the lost city of stilts and half. Philippine Star. https://www.philstar.com/metro/2001/09/19/134167/barangay-baseco-lost-city-stilts-and-halfAccessed 24 Sep 2024
Bless the Children Foundation (2024) Social impact and code of conduct. https://www.smokeytours.com/social-impact Accessed 24 Sep 2024
Cerojano T (2010) Manila slum fire leaves 4,000 homeless. CBS News. https://www.cbsnews.com/news/manila-slum-fire-leaves-4000-homeless/ Accessed 24 Sep 2024
Choi J (2021) Life in the peripheries: Tondo and Baseco. Johns Hopkins School of Advanced International Studies. http://www.saisperspectives.com/2021-22-issue/2021/12/27/life-in-the-peripheries-tondo-and-baseco Accessed 24 Sep 2024
Clapano J (2015) Mangrove farm to turn Baseco into tourist spot. The Philippine Star. p 17. https://ani.seafdec.org.ph/handle/20.500.12174/6363 Accessed 24 Sep 2024
Cutini V, Pinto VD, Rinaldi AM, Rossini F (2020) Proximal cities: does walkability drive informal settlements? Sustainability 12:756. https://www.mdpi.com/2071-1050/12/3/756 Accessed 24 Sep 2024
De Guzman C (2022) A large scale housing development for informal settlers in Manila. https://scholarspace.manoa.hawaii.edu/server/api/core/bitstreams/d261efa8-d2f6-4555-8da0-f37dc9aeefa9/content Accessed 24 Sep 2024
De Leon S (2023) First lady launches LAB for ALL in Baseco. https://pia.gov.ph/news/2023/07/26/first-lady-launches-lab-for-all-in-baseco Accessed 24 Sep 2024
Delizo M (2023) Mga bahay sa Baseco, nasunog. https://news.abs-cbn.com/news/12/13/23/mga-bahay-sa-baseco-nasunog Accessed 24 Sep 2024
Department of Environment and Natural Resources (DENR) (2020) DENR Chief leads planting of nilad, bakauan, nipa in Baseco. DENR News Alert, p 1. https://www.denr.gov.ph/images/DENR_News_Alerts/DENR_News_Alerts_26_December_2020_Saturday.pdf Accessed 24 Sep 2024
Dodds K, della Dora V (2018) Artificial islands and islophilia. In: Baldacchino G (eds) The Routledge international handbook of island studies: a world of islands. Routledge, London, pp 392–415
Feng Y, Hwang YH (2022) Mangroves on a landfill. Places J https://placesjournal.org/article/informal-settlers-environmental-rehabilitation-manila/?cn-reloaded=1 Accessed 24 Sep 2024
Filart G (2021) Implications and intentions: a day in Manila's biggest slum. https://www.ourworldinwords.com/implications-and-intentions-day-in-manilas-biggest-slum/ Accessed 24 Sep 2024
Habitat for Humanity (2007) HFH Philippines continues to build medium-rise homes for urban dwellers in Manila. https://www.habitat.org/ap/about/newsroom/2007-05-16-Philippines-Continues-To-Build-Medium-Rise-Homes Accessed 24 Sep 2024
Hrvacevic Z (2022) Dredging the mouth of Baseco lagoon. https://www.dredgingtoday.com/2022/05/02/dredging-the-mouth-of-baseco-lagoon/ Accessed 24 Sep 2024
International Alert (2019) Muslim mothers of Baseco launch social enterprise to build peace. https://www.international-alert.org/stories/muslim-mothers-baseco-launch-social-enterprise-build-peace/ Accessed 24 Sep 2024
Jackson M, della Dora,V (2011) Dreams so big only the sea can hold them: man-made islands as anxious spaces, cultural icons and travelling visions. Environ Plan A 41(9):2086–2104. https://doi.org/10.1068/a41237. Accessed 24 Sep 2024
Mayuga J (2020) What it takes to save Baseco Beach. Business Mirror. https://businessmirror.com.ph/2020/02/24/what-it-takes-to-save-baseco-beach/ Accessed 24 Sep 2024
Melican N (2015) Bayside residents ready for storm surges. Philippine Daily Inquirer. https://newsinfo.inquirer.net/706286/bayside-residents-ready-for-storm-surges Accessed 6 Sep 2024
Mendoza JE (2021) Manila Gov't awards 229 townhouse-style units to indigents in Baseco. Philippine Daily Inquirer. https://newsinfo.inquirer.net/1455664/manila-govt-awards-229-townhouse-style-units-to-indigents-in-baseco#ixzz7wNbjArQR Accessed 24 Sep 2024
Moaje M (2021) President Cory hospital to rise in Manila's Baseco compound. Philippine News Agency. https://www.pna.gov.ph/articles/1156975 Accessed 24 Sep 2024

Murphy D (2012) Baseco and its proclamation. https://www.ombudsman.gov.ph/UNDP4/wp-content/uploads/2012/12/BASECO_PROC1_HTML.pdf Accessed 24 Sep 2024
Ong G (2023) Dolomite, Baseco beaches unsafe for swimming. https://www.philstar.com/nation/2023/04/02/2256207/dolomite-baseco-beaches-unsafe-swimming Accessed 24 Sep 2024
PhilAtlas (2024) Barangay 649. https://www.philatlas.com/luzon/ncr/manila/barangay-649.html. Accessed 24 Sep 2024
Philippines Statistical Authority (2020) Census of population and housing. https://psa.gov.ph/content/2020-census-population-and-housing-2020-cph-population-counts-declared-official-president. Accessed 24 Sep 2024
Pilmico (2018) Pilmico turns over 7,000 kilos of rice in Baseco. https://www.pilmico.com/pilmico-turn-over-rice-to-baseco/ Accessed 24 Sep 2024
Prado J (2022) Caritas Manila builds material recovery facility for residents of Baseco. https://rcam.org/caritas-manila-builds-material-recovery-facility-for-residents-of-baseco/ Accessed 24 Sep 2024
Promchertchoo P (2019) Kidney for sale: inside Philippines illegal organ trade. Channel News Asia. https://www.channelnewsasia.com/asia/kidney-for-sale-philippines-illegal-organ-trade-857551 Accessed 24 Sep 2024
Rotary (2020) Baseco community garden project. https://map.rotary.org/en/project/pages/project_detail.aspx?guid=89ab4533-2375-4627-9e1d-acec0d1990c2 Accessed 24 Sep 2024
Samaniego A (2023) Xiaomi and world vision donate smart devices to Baseco high school. https://mb.com.ph/2023/12/31/xiaomi-and-world-vision-donate-smart-devices-to-baseco-high-school Accessed 24 Sep 2024
Shi H, Gao C, Dong C, Xia C, Xu G (2017) Variation of river islands around a large city along the Yangtze River from satellite remote sensing images. Sensors 17(10): 2213. https://www.mdpi.com/1424-8220/17/10/2213/pdf Accessed 24 Sep 2024
Smith L (2019) Baseco: Strategic pathways to formalize the informal. https://lup.lub.lu.se/luur/download?func=downloadFile&recordOId=8993255&fileOId=8993269. Accessed 24 Sep 2024
Soriano Z (2003) Countdown begins for Bagung Lupa's settlers. https://www.bulatlat.com/news/3-34/3-34-bagonglupa2.html Accessed 24 Sep 2024
Sreekumar K, Hassan G (2020) A study of artificial islands. Centre for Public Policy Research, Kochi. Accessed 24 Sep 2024
Tiangco MK (2020).Baseco gets access to clean water supply with 2 new water treatment systems. https://mb.com.ph/2020/11/11/baseco-gets-access-to-clean-water-supply-with-2-new-water-treatment-systems/ Accessed 24 Sep 2024
UNDESA (2019) Safe and prosperous BASECO. World vision urban programming. https://www.worldvision.org.ph/wp-content/uploads/2021/03/Safe-and-Prosperous-BASECO.pdf Accessed 24 Sep 2024
Urban Poor Associates (2022) Programs and projects. Brochure, Quezon City, Philippines
Valenzuela VPB, Esteban M, Onuki M (2020) Perception of disasters and land reclamation in an informal settlement on reclaimed land: Case of the BASECO Compound, Manila, The Philippines. Int J Disast Risk Sci 11(640–654):2024. https://doi.org/10.1007/s13753-020-00300-y. Accessed24September
Village Connect (2021) DENR Chief leads planting of Nilad, Bakauan, Nipa in Baseco. https://www.villageconnect.com.ph/denr-chief-leads-planting-of-nilad-bakauan-nipa-in-baseco/3155. Accessed 24 Sep 2024
Watanabe A (2007) The formation of migrant Muslim communities in Metro Manila. Kasarinlan 22(2):68–96. https://www.journals.upd.edu.ph/index.php/kasarinlan/article/view/695 Accessed 24 Sep 2024
World Vision (2019).More than 70 MT of plastic containers in Baseco through Waste to Wages Project. https://www.worldvision.org.ph/baseco-waste-to-wages-project-update/ Accessed 24 Sep 2024

World Vision (2020) World Vision provides aid for families displaced due to fire in Baseco. https://www.worldvision.org.ph/world-vision-provides-aid-for-families-displaced-due-to-fire-in-baseco/ Accessed 24 Sep 2024

Interviews (Respondents have freely waived their right to anonymity)

Diana Espinosa, Chairwoman of Barangay 649, Manila City. Interview on March 29, 2023

Edmundo Cayanan, Sangguniang Barangay Member of Barangay 649, Manila City. Interview on March 29, 2023

Ian Pido, Resident of Barangay 649, Manila City. Interview on February 5, 2024

Norberto Papa, Resident of Barangay 649, Manila City. Interview on February 5, 2024

Rolie Roldan, Resident of Barangay 649, Manila City. Interview on February 5, 2024

From Island to Island: Sri Lankans' Settlement Model(s) in the Three Main Cities of Sicily, Southern Italy

Francesca Bitonti, Federico Benassi, Angelo Mazza, Salvatore Strozza, and Luigi Scrofani

Abstract The so-called "Sri Lankan Diaspora in Italy" is the result of a migratory flow historically linking Italy and Sri Lanka. Indeed, Sri Lankans represent a long-established foreign community in Italy, with a notable presence in Southern Italy, particularly on the island of Sicily. Since large cities are Sri Lankans' preferred locations for settlement, this chapter offers a preliminary comparative spatial analysis of the settlement behaviors of Sri Lankans in the primary urban centers of Sicily: Palermo, Messina, and Catania. The work aims firstly to compare the settlement patterns of Sri Lankans across different urban contexts, assessing whether specificities prevail over macro-level similarities. Secondly, it attempts to detect possible spatial concentration of Sri Lankans at the intra-urban level and to verify spatial correlation with other variables proxies of socioeconomic heterogeneities in urban neighbourhoods. Multiple linear models have been run to assess the city-level influence of several socioeconomic predictors on the Sri Lankans' distribution. Moreover, geographically weighted regressions are implemented to assess the spatial dependence between Sri Lankans' location quotients and the covariates at the local level. Each spatial covariate has been interpolated into a unique geographic reference grid, in order to harmonise different types of areal unit and to perform comparisons between cities. Findings suggest that, despite moderate overall segregation, Sri Lankans concentrate in central affluent areas, revealing complex social polarization processes. Influenced by socioeconomic and demographic dynamics, their distribution reflects ethnic and migratory dimensions. While Sri Lankans gravitate towards wealthier Italian households, they also settle where foreign presence rises, indicating evolving ethnic segregation dynamics possibly influenced by gentrification and demographic shifts.

F. Bitonti (✉)
Department of Political and Social Sciences, University of Catania, Catania, Italy
e-mail: francesca.bitonti@unict.it

A. Mazza · L. Scrofani
Department of Economics and Business, University of Catania, Catania, Italy

F. Benassi · S. Strozza
Department of Political Sciences, University of Naples Federico II, Naples, Italy

T. Yamazaki and G. Baldacchino (eds.), *Relational Geographies of Islands in the Indo-Pacific*, International Perspectives in Geography 23,
https://doi.org/10.1007/978-981-96-7155-7_4

Keywords Island-to-island migration · Local-scale analysis · Segregation · Sicily · Sri Lankans

1 Introduction

Residential segregation, defined as the spatial separation of social groups within urban areas, arises from complex processes involving economic factors, individual choices, public policies, and social exclusion. It manifests as an unequal distribution of populations across urban spaces, reflecting significant disparities in access to resources and opportunities (Massey and Denton 1993). Socioeconomic segregation and inequality in urban areas have emerged as pressing global issues (van Ham 2021). Migration significantly affects and shapes both the social and spatial configurations of contemporary urban and metropolitan contexts, and the demographic composition of the host societies (Benassi et al. 2020a; Portes 2000; Strozza et al. 2016). Over the last three decades, Europe has become the arrival point of sustained and increasing foreign influxes, with the southern countries representing the areas predominantly affected by the phenomenon. These demographic shifts, compounded by the widening social and economic disparities resulting from the 2008 financial crisis and the 2020–2022 COVID-19 outbreak, have entrenched patterns of residential, economic, and social segregation in many European urban areas (Allen et al. 2004; Musterd et al. 2017; Tammaru et al. 2016). Residential segregation in Europe exhibits significant intra-regional variations, shaped by historical, economic, and institutional factors. In southern Europe, residential patterns are influenced by relatively recent immigration waves that intensified during the 1990s, resulting in migrant settlements in suburban or low-cost urban areas characterized by dense housing and affordable rents (Benassi et al. 2020). This contrasts sharply with northern and central Europe, where immigration began earlier, driven by post-war labor recruitment programs such as the Gastarbeiter policies of the 1960s in Germany. Consequently, northern European cities often display more structured ethnic segregation, facilitated by regulated housing markets and robust welfare systems (Maloutas 2016). Moreover, the predominance of private homeownership in southern Europe limits migrant access to public housing, compelling many to rely on substandard private rental markets (Arbaci 2007; Benassi et al. 2020). In contrast, northern and central European cities, with well-developed public housing sectors, often concentrate migrant populations in specific neighborhoods, creating more visible patterns of spatial segregation. These differences underscore the impact of institutional frameworks and housing policies on the socio-spatial integration of migrant communities within Europe.

In Italy, residential segregation related to migration is especially evident in metropolitan regions, which encompass extremely populated urban cores and their surrounding, less populated peripheries. In such contexts, demographic shifts and speculative real estate practices have intensified gentrification and hastened the suburbanization of poverty, further entrenching spatial economic separation (Hochstenbach and Musterd 2018). Although cities are widely regarded as engines of opportunity and economic development (OECD 2018a), many of them fail to achieve inclusive growth. Instead, they remain characterized by areas where poverty is concentrated, perpetuating existing social disparities and worsening the overall life conditions (OECD 2018a).

Socioeconomic disparities within urban areas are closely intertwined with the presence of foreign communities and their achievement of socialization and integration with the local populations. Migrants, who frequently settle in major cities and metropolitan regions, often face barriers such as language differences, religious beliefs, and the challenges of adapting to host societies (Vaughan and Arbaci 2011; Williamson 1988; Yap 1977). These obstacles make migrant groups particularly vulnerable to economic hardship and social exclusion, exacerbating existing urban inequalities (Dangschat 2016). In southern Europe, weaker economic conditions correlate with higher levels of residential segregation. Regions with less robust labour markets, lower productivity, and limited wealth tend to exhibit more pronounced spatial divisions (see: Benassi et al. 2020a, Benassi et al. 2020b). Understanding the local-level dynamics of residential segregation and spatial inequalities is critical in addressing disparities, marginalization, and social disruption. This is particularly relevant when considering the geographic distribution of migrant groups, such as Sri Lankans, in efforts to mitigate local urban disparities (OECD 2018b).

While research on this topic in Italy is growing, there is a scarcity of comparative analisyes among southern cities, particularly with a focus on specific immigrant communities. This is particularly surprising given the presence of immigrant groups, like Sri Lankans, who are heavily concentrated in certain urban areas of the country (Henayaka-Lochbihler and Lambusta 2004; Strozza et al. 2016). The historical connection between Sri Lanka and Italy primarily revolved around religion, with the initial wave of migration in the late 1970s consisting mainly of Catholic women seeking employment as caregivers to the elderly in the Italian whealthy families. In the long run, Sri Lankans have built strong local social ties, which now define the social networks of entire neighborhoods (Henayaka-Lochbihler and Lambusta 2004; Pathirage and Collyer 2011).

This chapter builds upon earlier works conducted by the authors (Benassi et al. 2023; Bitonti et al. 2023a) with the goal of providing a general understanding of Sri Lankans' settlement model in southern Italian cities. Here, the investigation is focused on their presence in Sicily's three metropolitan areas—Palermo, Messina, and Catania—and considers several explanatory variables and spatial statistical methods to achieve a clearer snapshot of the spatial allocation of the considered group. The chapter is structured as follows: the second section outlines the historical and geographical background of Sri Lankan immigration to Italy; the third section details the data and methods employed; the fourth section presents and discusses the results of the analysis; and the final section provides a summary and conclusion of the application.

2 Island to Island Migration and Sri Lankans in Italy

Islands can be examined, both as distinct geographical entities and as environments where the interactions between the island's ecosystem, developmental history, issues of progress and lack thereof, population dynamics, and migration can be observed with greater clarity under relatively controlled circumstances (King 2009). "Islands" and "migration" are inevitably intertwined. While an island is a clearly defined geographic feature—a landmass surrounded by water, typically the sea—migration represents the movement of people across space, a process that encompasses various forms of mobility. Many islands have a rich history of emigration, resulting in population loss through permanent departure, temporary relocation with eventual return, circular migration, or age-selective migration (Connell and King 1999). Emigrants from islands disperse to neighbouring islands, nearby landmasses, or far-flung countries and continents. Understanding island communities in the 21st century is often quite impossible without also looking carefully at their extensive diasporas (Olwig 1996, 2007). The development of most island states and territories has seen them missing out on the wave of industrialisation and urbanisation that gripped the planet over the last two centuries, heralding an exodus of the brightest, ablest, most mobile and most desperate. To the extent that, on the smallest island communities, the size of their diasporas may be larger than the resident population 'left' at home. Powerful circulations—of people, goods, ideas and financial capital—move between these groupings. The remittances of migrants, and their regular visits to the homeland as tourists, sustain island economies around the globe (Connell and Conway 2000). Conversely, islands have also attracted immigrants through relocation resulting from conflicts or the hope for better living conditions, and more recently, affluent individuals seeking peaceful retreats—a modern form of "conquest" reflecting the evolving global dynamics. Consequently, islands and their surrounding regions have progressively transformed into hubs of both national and international movements, where individuals and households navigate across diverse worlds and places, engaging in a dynamic interplay of multifaceted interpretations (Connell and King 1999).

The topic of islands and migrations is a subject of attention and study. In Italy, it has garnered consideration as a phenomenon related to emigration from the islands, especially among working-age Italians leaving the southern Italian islands, generally economically underdeveloped and characterized by high unemployment rates (Asso 2021; Bonifazi and Heins 2000). Less attention has gained the phenomenon of migration processes directed *towards* the Italian islands, such as Sicily, to the extent that they have become internationally attractive destinations for selected foreign nationalities.

The migration of Sri Lankans towards Italy traces its origins to the 1970s, with a marked increase during the 1980s (Henayaka-Lochbihler and Lambusta 2004). According to residence permit data, at the end of 1991, over 12,000 Sri Lankans were legally residing in Italy, 69% of whom were men. Notably, during the early 1990s, many Sri Lankans arrived in Italy illegally, mostly entering through the Balkan corridor (Morlicchio 1992). The 2001 Population and Housing Census reported that the Sri Lankan population had grown to over 26,000 individuals, in the whole country. Between 2002 and 2019, influx levels surged, with over 88,000 arrivals and fewer than 7,000 departures recorded. The sex ratio of the group has undergone fluctuations over time: the female dominated community chracterising the 1970s transitioned to a predominantly male population in the 1990s. More recently, a balanced gender distribution has emerged, with both males and females predominantly working as domestics (Näre 2012). Despite the association of domestic labor with women in Italy, Sri Lankan men also participate in this sector (Mazza et al. 2016).

The Sri Lankans' settlement model across Italy exhibits a unique dual characteristic. First, they display a preference for settling in major urban centres, where nearly 55% of the whole Sri Lankan population resided at the start of 2021 (Table 1). Big cities often have existing ethnic communities and networks that can provide support and assistance to new immigrants. These networks can include cultural organizations, religious institutions, and social groups that help newcomers integrate into society. Moreover, typically large urban areas have more extensive social services and government support systems than middle towns and rural villages. This can be crucial for immigrants who may need assistance with housing, language classes, or healthcare (Viñuela 2022). Second, they prefer to settle in southern cities—such as Messina, Palermo, and Catania in Sicily (see Fig. 1)—a trend seldom observed among other immigrant groups in Italy (Benassi et al. 2023) (see the bold numbers in the fourth and eighth columns of Table 1). Several Sri Lankans choose to reside in southern Italian cities due to the perceived similarities between the lifestyle in these urban areas and that of their home country (Henayaka-Lochbihler and Lambusta 2004).

The most plausible explanation suggests that migratory movements linking Sicily and Sri Lanka is largely driven by the transnational networks of Catholic institutions

Table 1 Sri Lankans residing in the main Italian municipalities (2012–2021)

Municipality (2012)	Sri Lankan residents (000)	%	% of foreigners	Municipality (2021)	Sri Lankan residents (000)	%	% of foreigners
Milan	11.1	15.5	6.3	Milan	16.1	14.8	6.3
Naples	7.2	10.1	22.9	Naples	15.3	14.1	26.3
Verona	5.2	7.3	17.4	Rome	9.1	8.4	2.7
Rome	4.9	6.8	2.2	Verona	7.4	6.8	20.1
Messina	3.6	5.0	**32.1**	Messina	3.8	3.5	**32.2**
Palermo	3.3	4.6	**16.7**	Palermo	3.1	2.9	**13.1**
Florence	1.6	2.2	3.7	Catania	2.4	2.2	**18.3**
Catania	1.4	2.0	**20.8**	Florence	2.1	1.9	4.2
Total	38.3	**53.5**		Total	59.3	**54.6**	

Source Italian General Population and Housing Census and selected town Population Registers

Fig. 1 Italy divided by regions with the evidence of Sicily (in dark yellow) and its three largest cities of Catania, Messina and Palermo (in red)

(Bacciocchi, 2010; Näre, 2008). This argument is reinforced by the strong connection between the majority of Sinhalese Sri Lankan immigrants (who are mainly Catholics) with religious symbolism, particularly evident in Sicily. The presence of pilgrimage sites and revered saints such as St. Anthony of Padua, the Tindari Madonna in Messina, St. Rosalia in Palermo (celebrated also by the local Hindu Tamil group, which is the largest Sri Lankan community of Tamil ethnicity located in Italy), and St. Agatha in Catania enhances Sicily's appeal to Sri Lankan immigrants. Additionally, these sacred sites serve to inspire migration aspirations among their relatives and friends in Sri Lanka (Benadusi 2015). In this sense, the impact of past migrations to Sicily, as described by Massey's concept of "cumulative causation" (1990), extends beyond just economic prosperity and material welfare.

Table 2 Selected variables and data sources

Variable	Source
Sri Lankans' location quotients	Italian population and housing census 2011
Share of large households (5+)	"
Share of foreigners over total population	"
Foreigners' mean growth rate (2001–2011)	"
Italians' mean growth rate (2001–2011)	"
Low educational attainment	"
Unemployment rate	"
Proportion of households living on rental property	"
Proportion of buildings in poor conditions	"
Average rental price per squared metre for private properties	OMI database 2011

3 Data and Methodology

The analysis of population data for Italians and Sri Lankans[1] is based on the 2011 Italian Population and Housing Census. However, it is important to acknowledge the under-coverage issues identified by the 2012 Post Enumeration Survey, which estimated an undercount of 1.1% for the total population and 11.1% for foreign nationals at the country level (Istat 2015). In order to implement the statistical analysis, additional geographically referenced covariates were employed to delineate the socioeconomic conditions of the urban areas where Sri Lankans reside. The main sources for this data, as listed in Table 2, include the 2011 Population and Housing Census and the Italian Revenue Agency's OMI database. The latter database was used to retrieve the average rental costs per square meter for private housing properties in 2011. All the analyses were conducted using the R statistical computing environment (R Core Team 2021).

The dataset analysed in the following application refers to specific geographic units, namely the census sections and the OMI sections, which differ in their design due to their distinct purposes. To address the modifiable areal unit problem (Openshaw 1984; Openshaw and Taylor 1979)—according to which the results of spatial analysis are influenced by the scale and type of geographical unit—we adopt an areal weighted interpolation method. This technique, commonly used for analysing socioeconomic data, involves transferring data from one set of spatial units (referred to as "source") to another set (the "target") (Goodchild et al. 1993). In this case, data from census sections and OMI zones were redistributed to a regular spatial grid of 100-by-100-m squares using the *aw_interpolate()* function from the "areal"

[1] The Sri Lankans data exploited for the present application were provided as part of the '*Caratteristiche, comportamenti e condizioni di vita degli immigrati di prima e di seconda generazione secondo le principali fonti disponibili*' research agreement between the Italian National Institute for Statistics (Istat), the National Research Centre (CNR) and six Italian universities.

R library (Prener and Revord 2019). The choice of this type of spatial grid, while inherently arbitrary, has become a widely accepted standard in recent studies on residential segregation and migrant settlement models (Data for Integration 2021). This standard is exemplified by the D4I Data Challenge on "Integration of Migrants in Cities" sponsored by the European Commission Joint Research Centre (JRC), which provided 100-m grids for population counts across urban areas in eight EU member states.[2] Numerous research studies have effectively leveraged the D4I grid to draw comparable conclusions about migrant residential dynamics in diverse European cities (Benassi et al. 2020a, b; Marcińczak et al. 2021; Olteanu et al. 2020).

To measure the overall level of segregation among Sri Lankans in the three urban areas studied, we calculated the corrected dissimilarity index (DI) (Mazza and Punzo 2015). A DI value below 0.30 suggests low segregation, values between 0.30 and 0.60 signify mild segregation, and values exceeding 0.60 reflect high segregation, though these thresholds may differ across national and local contexts (Massey and Denton 1993). To further examine local patterns of concentration or dispersion in the spatial allocation of Sri Lankans, we create geographic visualizations of their location quotients (LQs) (Haig 1926).

Earlier studies have underscored the phenomenon of ethnic occupational specialization in Palermo (Busetta et al. 2015), Messina (Bitonti et al. 2023b) and Catania (Mazza and Punzo 2016), with a predominant focus on domestic work within Italian households. This specialization has led to different neighbourhoods within these cities exhibiting varying levels of attractiveness to Sri Lankans, primarily driven by the availability of job opportunities. For this reason, we run different global and local regression analyses including several socioeconomic variables as possible contextual predictors for Sri Lankans' residential choices. First, in order to comprehensively and initially examine how Sri Lankans locate, we conducted multiple linear regressions (OLS) for each city. Then, to accurately capture the variations in the data at a local level and explicitly incorporate the spatial relationships within Sri Lankans' settlement model, we employed geographically weighted regressions (GWRs) (Fotheringham et al. 2002). Unlike OLS models, which assume uniform relationships across all geographic areas, GWR techniques allow for the estimation of location-specific regression models, accommodating the spatial heterogeneity often occurring in real-world scenarios (Brunsdon et al. 1998, 1999; Fotheringham et al. 2002). GWR technique enables the identification of spatially variable relationships between predictors and outcomes, uncovering localized trends that global analyses might overlook. Additionally, GWR models allow local parameters (local beta coefficients) to be mapped; this is a major advantage of local regression approaches (Matthews and Yang 2012).

[2] The European Commission's Knowledge Centre on Migration and Demography has recently launched the D4I dataset. This resource was developed through the spatial disaggregation of data derived from the 2011 Population and Housing Census, provided by national statistical institutes. The original data were processed to create a standardized grid, reporting the distribution of migrants within 100-by-100-m squares across all the cities of the following European countries: France, Germany, Ireland, Italy, the Netherlands, Portugal, Spain, and the United Kingdom.

4 Results

Our first aim is to assess if city-specific segregation dynamics emerge acroos the three cities considered. The bias-corrected DI values recorded for the three cities in 2011 indicate the level of residential segregation among the Sri Lankan communities in each municipality. Palermo has a relatively high DI of 0.66, suggesting a substantial portion of the Sri Lankan population is concentrated in specific neighborhoods or areas, leading to a notable separation from the Italians (the reference group). Messina's DI of 0.32 is notably lower than Palermo's, suggesting that Sri Lankans' distribution is more balanced with the Italian one, with less pronounced concentration in specific areas. Catania falls in between Palermo and Messina in terms of DI, with a value of 0.50. Its moderate level of residential segregation could indicate some degree of concentration in certain areas, but not as pronounced as in Palermo. Since the the global dissimilarity index (DI) yields different degrees of residential segregation at the urban level, we furhter investigate the possible presence of within-city variations in the group's spatial distribution in the three urban areas. Indeed, it is important to use both global and local indexes when measuring the settlement models of specific population groups (Brown and Chung 2006). To examine this, Fig. 2 illustrates the LQs for Sri Lankans in the three Sicilian cities. Initially, when examining the residential patterns within each urban context, a striking similarity emerges. Sri Lankans tend to be concentrated in distinct central neighbourhoods of the urban landscape. Notably, both in Messina, Palermo and Catania, the historical city centres and certain areas associated with affluent socio-economic profiles represent the areas with the largest presence of Sri Lankans. To some extent Messina presents a slightly larger areas of concentration up North. This intriguing observation warrants further exploration (see below). The uniformity in the distribution of Sri Lankans across these municipalities naturally leads one to consider the potential socioeconomic spatial factors that contribute to the observed residential patterns within these municipalities.

To assess which and how socioeconomic and demographic conditions affect the spatial distribution of the immigrant community, a non-spatial global multiple regression (OLS) and a spatial local multiple regression (GWR) have been carried out for each city. The results of the global models run for each city, as illustrated in Table 3, highlight three different types of relations between the dependent variable (Sri Lankans' location quotients) and the covariates. While the households on rent, the proportion of foreigners in 2001, and the foreigners' mean growth rate exhibit a positive association with LQs, the unemployment rate, the proportion of households composed by 5+ individuals, the proportion of individuals with a low educational attainment and the Italians' mean growth rate are negatively associated with the dependent variable in all the three cities. Conversely, the effect of the proportion of buildings in a poor conditions and the average rental price varies according to the urban context. Despite the statistical significance of these results, the overall performance of the global models proves to be very poor, as indicated by the low levels of

Fig. 2 Sri Lankans' LQs (reference group: global resident population) in the three largest cities of Sicily, 2011. Regular grid with 100 × 100 m squares. *Source* Authors' elaboration on Italian General Population and Housing Census data

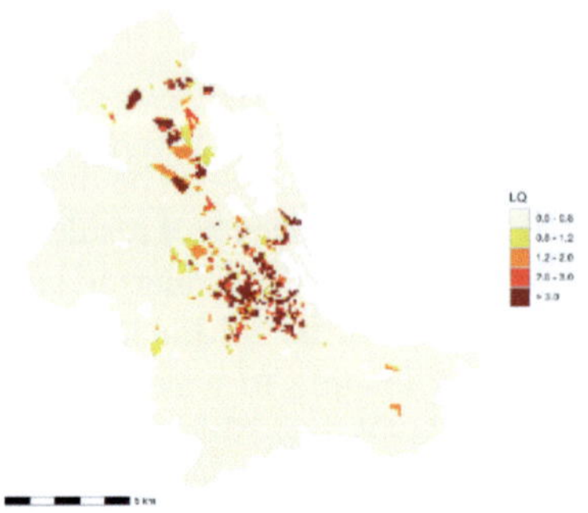

(a) Palermo

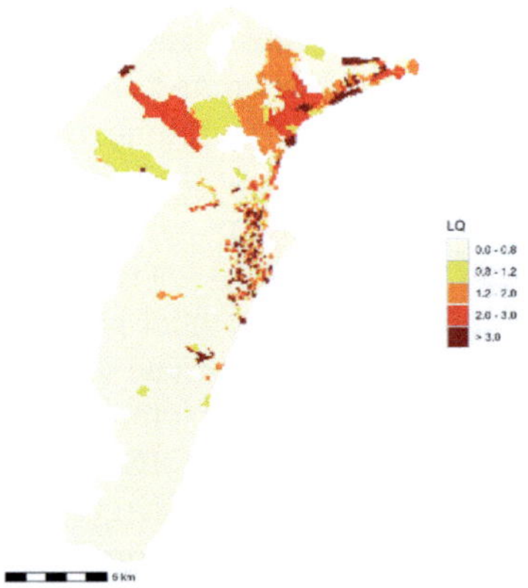

(b) Messina

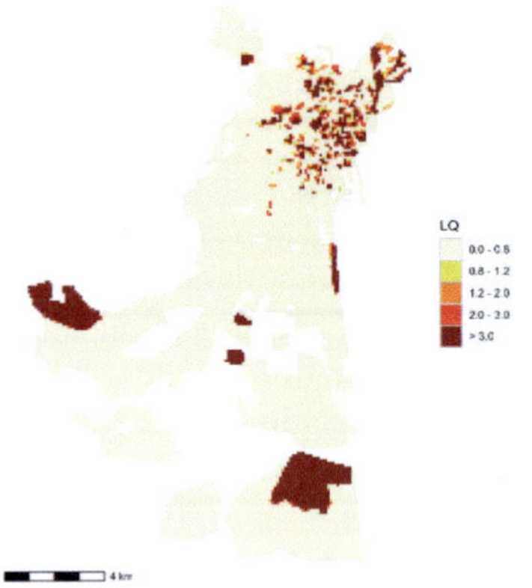

(c) Catania

Table 3 OLS coefficients and significance for three largest cities of Sicily, 2011, calculated on 100*100 m grids (* p value < 0.05; ** p value < 0.01; *** p value < 0.001)

Covariate	Palermo	Messina	Catania
(Intercept)	−0.080	1.672***	0.601**
Unemployment	−0.022***	−0.018***	−0.011
Large households	−0.052***	−0.036***	−0.023*
Households (living in rent)	0.049***	0.025***	0.012***
Low education	0.006	−0.027***	−0.030***
Foreigners' share (2001)	2.043***	13.404***	40.312***
Buildings (bad conditions)	−0.004***	−0.006***	0.009***
Average rental price	−0.022*	−0.098***	0.106***
Italians' growth rate	0.001	−0.003***	−0.004***
Foreigners' growth rate	0.001***	0.003***	0.003***
N° of cells	8,065	8,912	3,098
Adj. R^2	0.081	0.237	0.220

the adjusted R^2. This could indicate the presence of local heterogeneities in the relation between the outcome variable and its covariates which are not captured by the models implemented. Indeed, the spatial variations in some of the selected covariates illustrated in Fig. 3 show the presence of intra-urban variations in the socioeconomic conditions that could be accounted for implementing local analyses.

The implementation of a GWR for each city has widely improved the model's performance compared to the global analyses (see the AIC[3] and adjusted R^2 values reported in Table 4 which are, the first lower and the second considerably higher for GWR as compared to OLS). The variability of the phenomenon under investigation is also well captured, as demonstrated by the absence of spatial dependence in the residuals of the GWRs in Table 5, as opposed to the one detected under the OLS models. Specifically, the Moran's I test assessing the presence of (global) spatial dependence in the residuals of the model shows significant values very close to zero or non-significant values for the GWR residuals in both of the two selected neighbour configurations. Conversely, the OLS residuals appear to be positively and significatively correlated in all cases considered.

Considering the main summary statistics of the GWR coefficients for each of the three cities (Table 6), it is noticeable that the proportion of foreigners in 2001 consistently impacts Sri Lankans' LQs across cities. The variable exhibits high variability (in terms of standard deviation) and high proportions of significative coefficients in all the contexts analysed. Overall, the Sri Lankan presence seems to be affected by a demographic dynamic involving the residential choices of Italians and foreigners, as indicated by the high coefficients' standard deviations and significance of their mean

[3] The Akaike Information Criterion (AIC) is a statistical measure used to assess and compare the goodness of fit across different models. It strikes a balance between model complexity and its capacity to accurately explain the data (Akaike 1974).

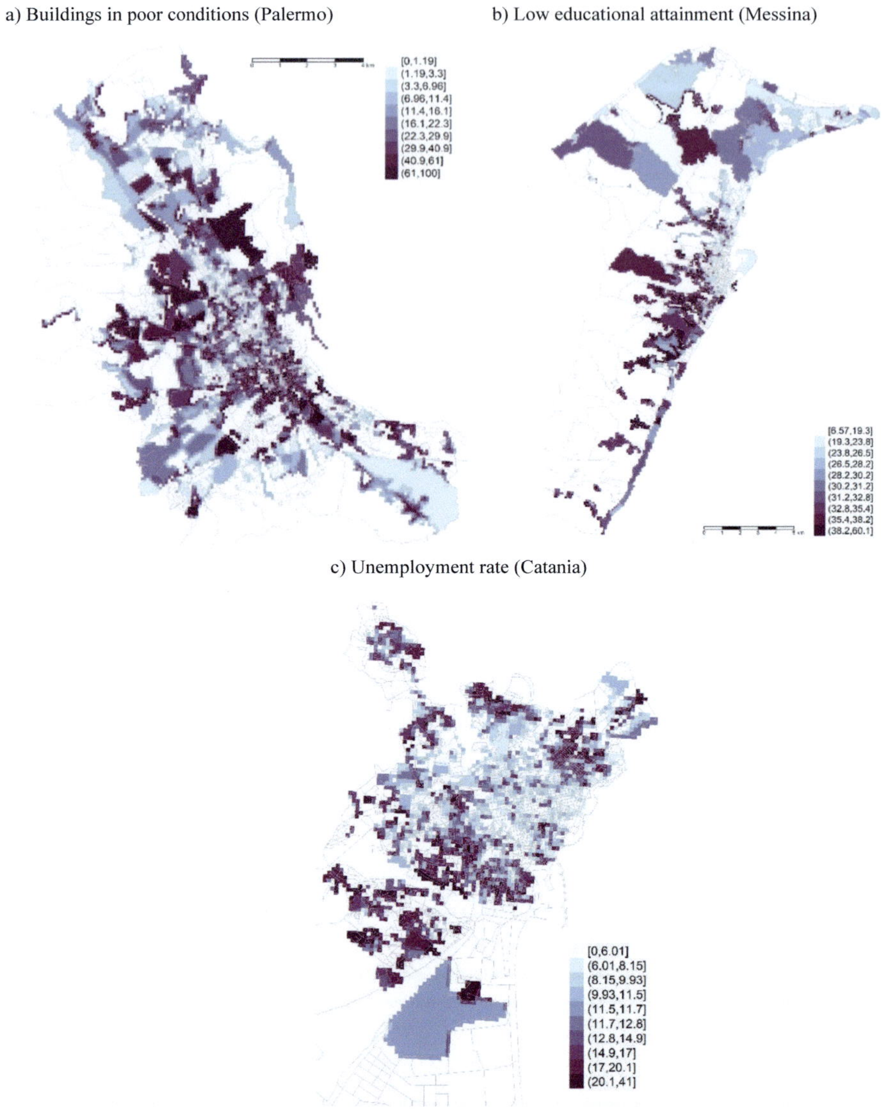

Fig. 3 Suburban distribution for selected independent variables in the three cities of Palermo, Messina and Catania. Time period: 2011. *Source* Authors' elaboration on Italian General Population and Housing Census data

growth rates. In terms of proportion of GWR significant coefficients, this influence seems to be stronger in Palermo and Catania than in Messina. Besides this similarity across cities, some peculiarity stands out: Palermo records a strong influence from mean rent cost, indicating potential housing challenges; in Messina the large standard

Table 4 OLS and GWR model performance for the three selected cities

Municipality	AIC		Adjusted R^2	
	OLS	GWR	OLS	GWR
Palermo	32,794.74	18,824.14	0.08	0.83
Messina	27,146.9	12,223.42	0.24	0.85
Catania	14,178.24	11,692.5	0.22	0.63

Table 5 Moran's I test for residual spatial dependence under different neighbours structure for OLS and GWR models (* p value < 0.05; ** p value < 0.01; *** p value < 0.001)

Municipality	N° of neighbours			
	30		50	
	OLS	GWR	OLS	GWR
Palermo	0.35***	−0.005	0.28***	−0.01***
Messina	0.41***	0.015***	0.36***	0.002
Catania	0.15***	0.009**	0.1***	−0.004

deviation and the high proportion of significative GWR coefficients for the households living in rent, could indicate a rental dynamics; finally, Catania exhibits unique influences from mean rent cost, households in rentals and proportion of building in bad conditions, pointing to intricate housing and economic relationships.

The GWR results highlight that the relationship between the location quotients of Sri Lankans and their correlates varies across the three municipalities studied. Some independent variables show a similar effect on the outcome variable (LQ) across urban areas, while other covariates have city-specific influence.

For the sake of brevity, Figs. 4 and 5 show the local coefficient estimates for just few selected independent variables (at the 5% significance level). For the covariates "buildings in poor conditions" and "average rental price", the distribution of coefficients is considerably similar across cities. In particular, all the three cities are characterised by a core-periphery duality, where the configuration in the urban centres appear more fragmented than in the peripheries. Catania stands out for its more marked and peculiar situation than in the other two cities. Indeed, in the peripheral and wealthy areas up north, the proportion of buildings in a poor state of maintenance and the mean rent for private properties are positively and negatively correlated with the Sri Lankans' LQs, respectively. Such dynamics could alternatively suggest that in those neighbourhoods many Sri Lankans live within their employers' household (as in the case of elderly assistance), or that they reside in some specific block generally inhabited by workers specialised in caregiving and housekeeping services usually requested by rich Italian households, the latter implying some sort of ethnic separation. Observing the values of the GWR coefficients referring to the mean rent cost in Catania, a sharp centre-periphery dichotomy arises. Specifically, the association between Sri Lankans' LQs and the covariate changes from positive

Table 6 GWR coefficients' summary statistics for the three selected cities (statistical significance: $\alpha = 0.05$)

Palermo					
Variable	Min	Max	Mean	Standard deviation	Significant coeffs. (%)
Intercept	−29.801	31.823	−0.544	2.71	20.547
Unemployment	−0.685	0.989	−0.010	0.12	18.537
Large households	−0.635	1.301	0.021	0.145	19.074
Households (living in rent)	−0.292	0.37	0.010	0.046	19.136
Low education	−0.412	0.451	0.006	0.072	17.476
Foreigners'share (2001)	−252.625	310.788	32.261	65.782	33.729
Buildings (bad conditions)	−0.117	0.115	−0.001	0.017	16.003
Average rental price*	−6.49	27.234	0.386	1.983	23.917
Italians' growth rate	−0.222	0.049	−0.005	0.019	15.004
Foreigners' growth rate	−0.012	0.02	0.001	0.003	20.821
Messina					
Variable	Min	Max	Mean	Standard deviation	Significant coeffs. (%)
Intercept	−22.678	64.342	1.853	8.929	41.309
Unemployment	−0.808	0.921	0.005	0.163	33.938
Large households	−3.848	5.247	0.007	0.273	40.322
Households (living in rent)	−7.899	2.263	−0.044	0.956	39.608
Low education	−2.183	1.052	−0.028	0.149	39.778
Foreigners' share (2001)	−338.456	440.225	9.801	42.079	23.211
Buildings (bad conditions)	−0.716	0.512	−0.011	0.111	37.657
Average rental price	−5.581	1.133	−0.096	0.560	15.693
Italians' growth rate	−0.026	0.009	−0.001	0.003	19.957
Foreigners' growth rate	−0.007	0.019	0.001	0.002	21.148

(continued)

Table 6 (continued)

Catania					
Variable	Min	Max	Mean	Standard deviation	Significant coeffs. (%)
Intercept	−16.275	30.507	0.459	4.677	27.726
Unemployment	−0.482	0.329	−0.023	0.096	26.399
Large households	−0.532	0.602	−0.037	0.132	28.761
Households (living in rent)	−0.085	0.297	0.039	0.071	35.62
Low education	−0.408	0.249	−0.010	0.08	25.137
Foreigners' share (2001)	−13.155	176.477	39.124	42.933	49.757
Buildings (bad conditions)	−0.083	0.221	0.015	0.053	25.687
Average rental price	−8.557	2.215	−0.306	0.978	25.073
Italians' growth rate	−0.071	0.017	−0.007	0.012	28.049
Foreigners' growth rate	−0.001	0.017	0.004	0.004	44.969

* *Note* The summary statistics referring to the "Mean rent cost" coefficients are computed on their trimmed distribution due to the presence of a very high outlier

to negative along the centre-to-periphery direction. This situation seems to suggest that the Sri Lankans' settlement model does not reflect the sole adaptation to the Italian employers' needs as reflected by the effect of the proportion of buildings in bad conditions, at least in this city. Examining the coefficients of the average growth rates for Italians and foreigners during 2001–2011, it becomes evident that in all the three cases, Sri Lankans are more likely to live in the urban centres where the share of foreigners has risen, while the share of Italians has declined over time. This trend points to a form of spatial separation between foreigners (including Sri Lankans) and Italians, which appears to transcend the specific characteristics of each city.

5 Discussion and Conclusions

Small worlds are teaming with global lives (King and Connell, 1999): islands, and small islands in particular, owe so much to other islands and mainland, from and with which they secure remittance payments, employment and educational opportunities, items of consumption, specialist labour, tourists and much more (King, 2009). It is practically impossible to fully understand small island life without acknowledging the role of sprawling island diasporas (e.g., Olwig and Sorensen, 2003).

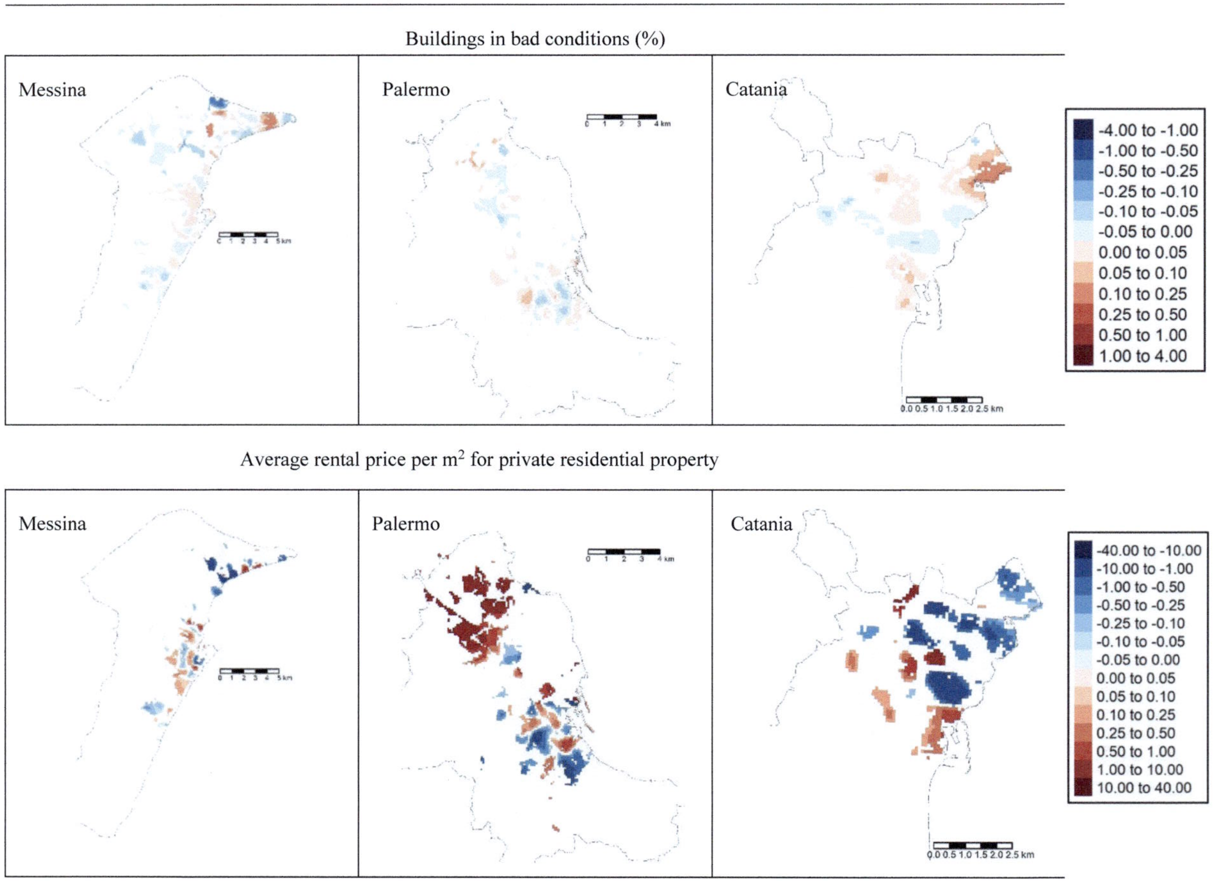

Fig. 4 GWR results: significant local coefficient estimates for building maintenance condition and mean rent cost per m^2 for residential properties in the Sicilian cities of Palermo, Messina, and Catania (5% significance level; geographical scales not comparable)

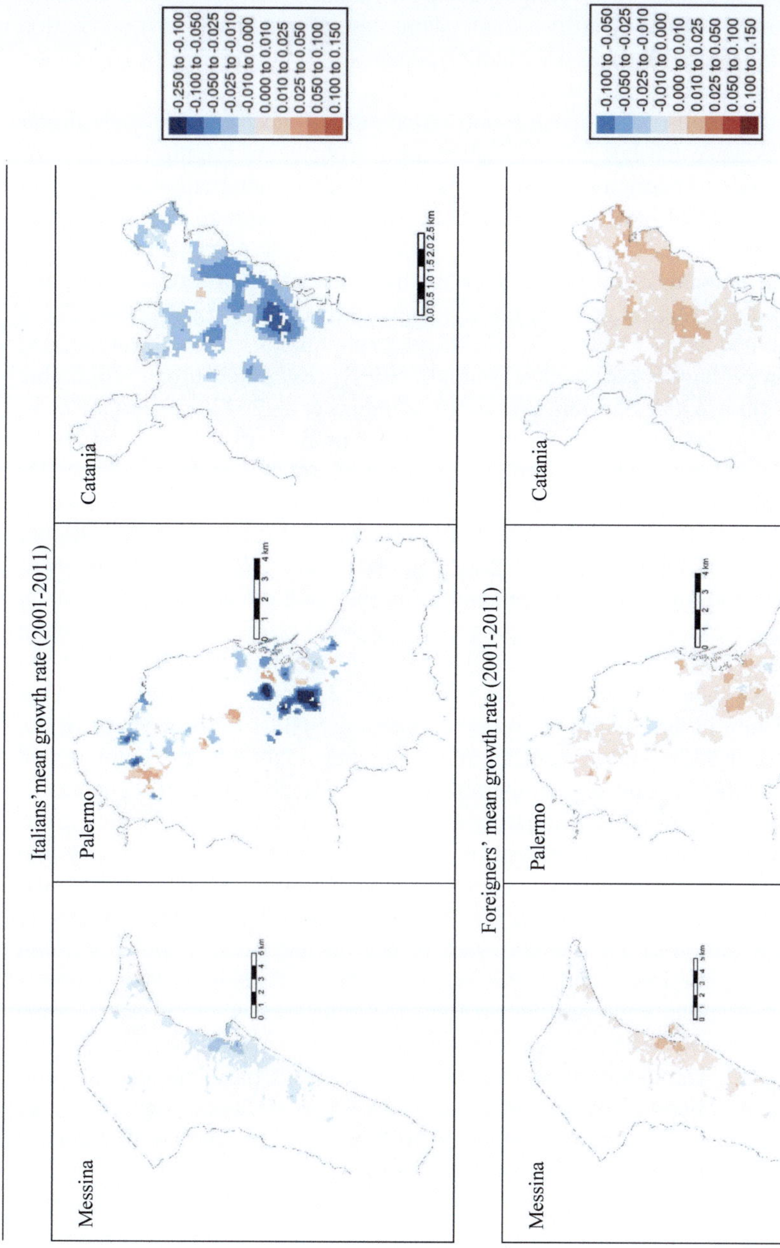

Fig. 5 GWR results: significant local coefficient estimates for Italians' and foreigners' mean growth rates (2001–2011) in the Sicilian cities of Palermo, Messina, and Catania (5% significance level; geographical scales not comparable)

This study may be one of the first to alert us to the extent to which island-to-island migration and settlement may be pertinent, and not just where *small* islands are concerned. Sri Lanka is a sovereign state in the Indian Ocean with a resident population of around 22 million; hardly a small island state. While Sicily, an island region of Italy, is the largest island in the Mediterranean Sea, in terms of both population (5 million) and land area (25,700 km^2).

Additionally, Southern European countries have experienced a shift in their migration dynamics since the late 1970s, transitioning from being traditionally labour-exporting nations to becoming destinations for significant international migratory flows (Collinson 1993; de Haas et al. 2020). Moreover, Southern European cities tend to be more dense compared to their Northern counterparts (Kasanko et al. 2006), and the housing sector is characterized by a substantial prevalence of home-ownership and a strong reliance on family assets (Arbaci 2019; Leal 2004). These factors, combined with a historically lower foreign population presence compared to Northern Europe, have contributed to Southern cities' greater intra-urban geographic variations (Arbaci 2008; Barbieri et al. 2018; Ciommi et al. 2022; López-Gay et al. 2020; Malheiros 2002). In European cities, poor living conditions (resulting in reduced disparity between local population and immigrant groups due to a generally worse economic condition), a highly informal real estate market, and a multifaceted spatial allocation of the population lead to lower residential segregation and a higher degree of suburbanization (Malheiros 2002). Furthermore, in Italy, the lack of regulation in urban planning and development during the rapid growth period following WWII may have played a role in shaping the complexity of residential arrangement in urban areas.

Our study is exploratory; yet its findings invite a fresh perspective on certain claims commonly found in the existing literature concerning Southern European contexts, residential segregation, and socioeconomic inequalities. Notably, the employment of spatial units that are entirely comparable through a standardized grid enables us to conduct comparisons while mitigating the influence of varying aggregation units. The conducted study aims to analyse the Sri Lankan community at the city and with-city level across three distinct urban reas, all of which are located in Sicily. Overall, the results of global indices and regressions have highlighted a moderate degree of segregation of Sri Lankans with respect to the Italian population, with Messina showing the lowest value. Nevertheless, the local analyses have shown a complex spatial configuration in the Sri Lankans' allocation with respect to some contextual covariates, although the overall pattern seems to be similar for all three cities considered. The unique characteristics of local residential markets can pose challenges in establishing overarching patterns on a metropolitan or regional scale. Nevertheless, the significant convergence of Sri Lankans in the central, particularly affluent, neighbourhoods of the cities highlights the existence of social polarization processes in act. These dynamics, as noted in other European urban contexts (Haandrikman et al. 2021; Martín-Legendre et al. 2021; Porcel and Antón-Alonso 2021), have a pronounced ethnic and migratory dimension (Boterman et al. 2021; Demireva and Zwysen 2021). In the case of the main Sicilian cities, a double dynamic seems to affect the localisation of Sri Lankans: first, Sri Lankans concentrate in the wealthiest neighbourhoods mostly inhabited by rich Italian households, second, they also live in areas where the foreign presence has increased, and the Italian one has decreased over time. In an attempt to interpret and synthetise these two influencing factors, one socioeconomic and one demographical, some considerations arise. According to the prevalence of one dynamic over the other, the effect on ethnic segregation in terms of repulsion/attraction between Italians and Sri Lankans could be heterogeneous over time. Moreover, the possible occurrence of some underlying gentrification processes or changes in the ethnic composition of neighbourhoods could shape urban-specific configurations, as have already partially emerged in Catania. In this sense, the findings yielded by the current application have laid the ground for future research aimed at broadly assessing the validity of such arguments and considerations.

The primary limitation of the present work lies in the failure to consider the different generations of Sri Lankans who arrived in the three cities. The lack of stratification between generations could have overcome meaningful differences in the allocation preferences of old and new immigrants. The consideration of the arrival time of each individual could represent an important improvement for the future extension of the current application, given the actual data availability.

Understanding the spatial distribution of migrant populations like the Sri Lankan community is essential for policymakers, who can use this information to develop targeted strategies to enhance the well-being and integration of Sri Lankan migrants in these cities. The results of this and similar studies can inform policy decisions related to housing, employment, education, and social integration with respect to various ethnic migrant communities.

Statements and Declarations

This study has been partially supported by the PRIN 2022-PNRR research project "Foreign population and territory: integration processes, demographic imbalances, challenges and opportunities for the social and economic sustainability of the different local contexts (For.Pop.Ter)" [P2022WNLM7], funded by European Union—Next Generation EU, component M4C2, Investment 1.1. It has also been partially supported by the Italian Ministerial grant PRIN 2022 "Depopulation Risk: Experiences, Mobility and Subjective Well-being (DREAMS)" [2022RNKSEL] CUP E53D23010380006. It has also been partially supported by the PNRR—Missione 4 "Istruzione e Ricerca"—Componente 2 "Dalla ricerca all'impresa"—Investimento 1.3—Partenariato Esteso "GRINS—Growing Resilient, INclusive and Sustainable", cod. PE0000018, CUP E63C22002120006, Spoke 8—"Social sustainability", finanziato dall'Unione Europea—NextGenerationEU.

Data Availability Statement

General population data analysed in this study were obtained from the Istat online repository and are available at: https://www.istat.it/it/archivio/104317. Data about the mean rent cost were retrieved from the online OMI repository and are available at: https://www1.agenziaentrate.gov.it/servizi/Consultazione/ricerca.htm.

References

Akaike H (1974) A new look at the statistical model identification. IEEE Trans Autom Control 19(6):716–723

Allen J, Barlow J, Leal J, Maloutas T, Padovani L (2004) Housing and welfare in southern Europe. Blackwell, Oxford

Arbaci S (2007) Ethnic segregation, housing systems and welfare regimes in Europe. Eur J Hous Policy 7(4):401–433

Arbaci S (2008) (Re)Viewing ethnic residential segregation in Southern European cities: housing and urban regimes as mechanism of mar-ginalisation. Hous Stud 23:589–613

Arbaci S (2019) Paradoxes of segregation: housing systems, welfare regimes and ethnic residential change in southern European cities. John Wiley and Sons, Oxford, UK

Asso PF (2021) New perspectives on old inequalities: Italy's north–south divide. Territ Polit Gov 9(3):346–364

Bacciocchi L (2010) Tra Sri Lanka e Italia. Vite, percorsi, mondi migranti e ambivalenze delle migrazioni. Italy: University of Verona

Barbieri GA, Benassi F, Mantuano M, Prisco MR (2018) In search of spatial justice. Towards a conceptual and operative framework for the analysis of inter-and intra-urban ine-qualities using a geo-demographic approach. The case of Italy. Reg Sci Policy Pract 11:109–121. https://doi.org/10.1111/rsp3.12158

Benadusi M (2015) Sacred spaces and border crossing: Sinhalese dreams of a Sri Lanka-Sicily round trip. Sri Lanka J Soc Sci 38(2):95–105

Benassi F, Bonifazi C, Heins F, Lipizzi F, Strozza S (2020a) Comparing residential segregation of migrant populations in selected European urban and metropolitan areas. Spat Demogr 8:269–290. https://doi.org/10.1007/s40980-020-00064-5

Benassi F, Pascual-Iglesias R, Salvati L (2020b) Residential segregation and social diversification: exploring spatial settlement patterns of foreign population in Southern European cities. Habitat Int 101(1022):00. https://doi.org/10.1016/j.habitatint.2020.102200

Benassi F, Bitonti F, Mazza A, Strozza S (2023) Sri Lankans' residential segregation and spatial inequalities in Southern Italy: an empirical analysis using fine-scale data on regular lattice geographies. Qual Quant 57(2):1629–1648. https://doi.org/10.1007/s11135-022-01434-5

Bitonti F, Benassi F, Mazza A, Strozza S (2023) From South Asia to Southern Europe: a comparative analysis of Sri Lankans' residential segregation in the main Italian cities using high-resolution data on regular lattice geographies. Genus 79:23. https://doi.org/10.1186/s41118-023-00203-y

Bitonti F, Mazza A, Mucciardi M, Scrofani L (2023b) Urban transformations and the spatial distribution of foreign immigrants in Messina. In: Brentari E, Chiodi M, Wit E (eds) Models for data analysis. Selected papers of 49th meeting of Italian statistical society, pp 53–67. Springer Proceedings in Mathematics and Statistics

Bonifazi C, Heins F (2000) Long-term trends of internal migration in Italy. Int J Popul Geogr 6(2):111–131

Boterman WR, Musterd S, Manting M (2021) Multiple dimensions of residential segregation. The case of the metropolitan area of Amsterdam. Urban Geogr 42:481–506. https://doi.org/10.1080/02723638.2020.1724439

Brown LA, Chung SY (2006) Spatial segregation, segregation indices and the geographical perspective. Popul Space Place 12(2):125–143

Brunsdon C, Fotheringham AS, Charlton ME (1998) Geographically weighted regression-modelling spatial non-stationarity. J R Stat Soc D 47(3):431–443

Brunsdon C, Fotheringham AS, Charlton ME (1999) Some notes on parametric significance tests for geographically weighted regression. J Reg Sci 39(3):497–524

Busetta A, Mazza A, Stranges M (2015) Residential segregation of foreigners: an analysis of the Italian city of Palermo. Genus 71(2–3):177–198. https://www.jstor.org/stable/genus.71.2-3.177. Accessed 5 Oct 2024

Ciommi M, Egidi G, Vardopoulos I, Chelli FM, Salvati L (2022) Toward a 'Migrant Trap'? Local development, urban sustainability, sociodemographic inequalities, and the economic decline in a Mediterranean metropolis. Soc Sci 12(1):26
Collinson S (1993) Europe and international migration. Pinter Publisher, London
Connell J, King R (1999) Island migration in a changing world. In: Connell J, King R (eds) Small worlds, global lives. Islands and migration, pp 1–26. London: A&C Black
Connell J, Conway D (2000) Migration and remittances in island microstates: a comparative perspective on the South Pacific and the Caribbean. Int J Urban Reg Res 24(1):52–78
Dangschat JS (2016) Residential segregation by nationality—a discourse full of contradictions. Österreichische Zeitschrift für Soziologie 41:81–101
Data for Integration (D4I) (n.d.) Retrieved 21 Sept 2021, https://knowledge4policy.ec.europa.eu/migration-demography/data-integration-d4i_en#factsheets. Accessed 5 Oct 2024
de Haas H, Castles S, Miller MJ (2020) The age of migration: international population movements in the modern world, 6th edn. Red Globe Press, London
Demireva N, Zwysen W (2021) Ethnic enclaves, economic and political threat: an investigation with the European social survey. Front Sociol 6:660378. https://doi.org/10.3389/fsoc.2021.660378
Fotheringham AS, Brunsdon C, Charlton ME (2002) Geographically weighted regression: the analysis of spatially varying relationships. Wiley, Chichester
Goodchild MF, Anselin L, Deichmann U (1993) A framework for the areal interpolation of socioeconomic data. Environ Plan A 25:383–397
Haandrikman K, Costa R, Malmberg B, Rogne AF, Sleutjes B (2021) Socio-economic segregation in European cities. A comparative study of Brussels, Copenhagen, Amsterdam, Oslo and Stockholm. Urban Geogr. https://doi.org/10.1080/02723638.2021.1959778.
Haig RM (1926) Toward an understanding of the metropolis. Quart J Econ 40:421–433
Henayaka-Lochbihler R, Lambusta M (2004) The Sri-Lankan Diaspora in Italy. http://citeseerx.ist.psu.edu/viewdoc/download?doi=10.1.1.564.2219&rep=rep1&type=pdf. Accessed 5 Oct 2024
Hochstenbach C, Musterd S (2018) Gentrification and the suburbanisation of poverty: changing urban geographies through boom-and-bust periods. Urban Geogr 39:26–53. https://doi.org/10.1080/02723638.2016.1276718
Istat (2015) Indagine di copertura del 15° Censimento generale della popolazione e delle Abitazioni
Kasanko M, Barredo JI, Lavalle C, McCormick N, Demicheli L, Sagris V, Brezger A (2006) Are European cities becoming dispersed? A comparative analysis of 15 European urban areas. Landsc Urban Plan 77(1–2):111–130
King R, Connell J (eds) (1999) Small worlds, global lives: Islands and migration. Pinter, London
King R (2009) Geography, islands and migration in an era of global mobility. Isl Stud J 4(1):53–84
Leal JM (2004) El diferente modelo residencial de los países del Sur de Europa: El mercado de viviendas, la familia y el estado. Arxius De Sociologia 10:11–37
López-Gay A, Andújar-Llosa A, Salvati L (2020) Residential mobility, gentrification and neighborhood change in Spanish cities: a post-crisis perspective. Spat Demogr 8(3):351–378
Malheiros JM (2002) Ethni-cities: residential patterns in the Northern European and Mediterranean metropolises—implications for policy design. Int J Popul Geogr 8(2):107–134
Maloutas T (2016) Introduction: Residential segregation in context. In: Fujita K (ed) Residential segregation in comparative perspective. Routledge, London, pp 1–36
Marcińczak S, Mooses V, Strömgren M, Tammaru T (2021) A comparative study of immigrant-native segregation at multiple spatial scales in urban Europe. J Ethn Migr Stud. https://doi.org/10.1080/1369183X.2021.2008887
Martín-Legendre JI, Castellanos-García P, Sánchez-Santos JM (2021) Neighbourhood inequality and spatial segregation: an analysis with tax data for 40 Spanish cities. Cities 118:103354. https://doi.org/10.1016/j.cities.2021.103354
Massey DS (1990) Social structure, household strategies and the cumulative causation of migration. Popul Index 56(1):3–26
Massey DS, Denton NA (1993) American apartheid: segregation and the making of the underclass. Mass. Harvard University Press, Cambridge

Matthews SA, Yang TC (2012) Mapping the results of local statistics: using geographically weighted regression. Demogr Res 26(6)
Mazza A, Punzo A (2015) On the upward bias of the dissimilarity index and its corrections. Sociol Methods Res 44:80–107
Mazza A, Punzo A (2016) Spatial attraction in migrants' settlement patterns in the city of Catania. Demogr Res 35:117–138
Morlicchio E (1992) I Singalesi e i Tamil. In: Mottura G (ed) L'Arcipelago Immigrazione: caratteristiche e modelli migratori dei lavoratori stranieri in Italia. Ediesse, Rome, pp 289–299
Musterd S, Marcińczak S, van Ham M, Tammaru T (2017) Socio-economic segregation in European capital cities: increasing separation between poor and rich. Urban Geogr 38(7):1062–1083
Näre L (2008) La comunità transnazionale dello Sri Lanka. In: Colombo A, Sciortino G (eds) Stranieri in Italia. Trent'anni dopo, pp 83–116. Bologna: Il Mulino
Näre L (2012) Migrancy, gender, and social class in domestic labour and social care in Italy: an intersectional analysis of demand. J Ethn Migr Stud 39:601–623
OECD (2018) Divided cities. Understanding Intra-Urban inequalities, Paris
OECD (2018b) Foreword of the book: divided cities. Understanding Intra-urban Inequalities. https://www.oecd-ilibrary.org/sites/9789264300385-en/index.html?itemId=/content/publication/9789264300385-en. Accessed 5 Oct 2024
Olteanu M, de Bezenac C, Clark W, Randon-Furling J (2020) Revealing multiscale segregation effects from fine-scale data: a case study of two communities in paris. Spat Demogr 8(3):255–267. https://doi.org/10.1007/s40980-020-00065-4
Olwig KF (1996) Global culture, island identity. Taylor and Francis
Olwig KF (2007) Caribbean journeys: an ethnography of migration and home in three family networks. Duke University Press
Olwig KF, Sorensen NN (2003) Work and migration: life and livelihoods in a globalizing world. Routledge
Openshaw S (1984) Ecological fallacies and the analysis of areal census data. Environ Plan A 16:17–31
Openshaw S, Taylor PJ (1979) A million or so correlation coefficients: three experiments on the modifiable areal unit problem. In: Wrigley N (ed) Statistical applications in spatial sciences. Pion, London, pp 127–144
Pathirage J, Collyer M (2011) Capitalizing social networks: Sri Lankan migration to Italy. Ethnography 12:315–333
Porcel S, Antón-Alonso F (2021) Barcelona y el modelo paradójico de segregación residencial. Encrucijadas: Revista Crítica de Ciências Sociais 21:a2103
Portes A (2000) Immigration and the metropolis: reflections on urban history. J Int Migr Integr/revue De L'integration Et De La Migration Internationale 1:153–175
Prener C, Revord C (2019) Areal: an R package for areal weighted interpolation. J Open Source Softw 4:1221. https://doi.org/10.21105/joss.01221
R Core Team (2021) R: a language and environment for statistical computing. Vienna, Austria: R Foundation for Statistical Computing
Strozza S, Benassi F, Ferrara R, Gallo G (2016) Recent demographic trends in the major Italian urban agglomerations: the role of foreigners. Spat Demogr 4:39–70
Tammaru T, Marcińczak S, van Ham M, Musterd S (2016) Socioeconomic segregation in European capital cities: East meets West (Regions and Cities). Routledge, London
van Ham M, Tammaru T, Ubarevičienė R, Janssen H (2021) Urban socio-economic segregation and income inequality. A global perspective. Springer, Berlin
Viñuela A (2022) Immigrants' spatial concentration: region or locality attractiveness? Popul Space Place 28(2):e2530
Williamson JG (1988) Migration and urbanization. Handb Dev Econ 1:425–465
Yap LY (1977) The attraction of cities: a review of the migration literature. J Dev Econ 4(3):239–264

Countermeasures to Climate Insecurity of Asia-Pacific Island Countries: Out-Migration and Adaptation Options

Juha I. Uitto, Mikiyasu Nakayama, Carl Bruch, Miko Maekawa, and Akiko Sakamoto

Abstract This chapter explores the challenges and dynamics associated with climate-induced migration from small island developing states (SIDS), focusing on atoll nations facing sea-level rise. Drawing upon existing studies, as well as evaluative evidence from on-the-ground initiatives, the chapter scrutinizes possible means to maintain peace and security in Pacific and Indian Ocean SIDS by enhancing their resilience to climate-induced risks and by nurturing sustainable resource management policies and practices. While migration is seen as an adaptation strategy, it presents risks and insecurities, necessitating enhanced attention to climate change adaptation. The study evaluates various adaptive measures, emphasizing the need for sustainable solutions to retain population and territory. It also examines the role of international organizations, particularly the Global Environment Facility (GEF), in supporting SIDS in building resilience and implementing integrated environmental management approaches. Two case studies—of the Marshallese in the United States and the I-Kiribati in Fiji—are presented to contextualize the arguments. The paper introduces the concept of the 'Migration with Dignity' Framework, rooted in existing human rights principles, as a comprehensive approach to address climate-induced migration challenges. The findings highlight the importance of considering the right to stay and propose recommendations for sustainable development, resilience-building, and supportive measures for climate migrants.

Keywords Small islands · Sustainable development · Adaptation · Climate change · Migration with dignity

---

J. I. Uitto (✉) · C. Bruch
Environmental Law Institute, Washington, D.C., USA
e-mail: uitto@eli.org

M. Nakayama · A. Sakamoto
Global Infrastructure Fund Research Foundation Japan, Tokyo, Japan

M. Maekawa
Sasakawa Peace Foundation, Tokyo, Japan

T. Yamazaki and G. Baldacchino (eds.), *Relational Geographies of Islands in the Indo-Pacific*, International Perspectives in Geography 23,
https://doi.org/10.1007/978-981-96-7155-7_5

1 Introduction

Small island countries face insecurity due to ongoing climate change, causing both loss of territory and massive outmigration. While outmigration can represent an adaptation measure to climate change, and thus prevent internal conflicts as land and water resources are lost due to sea level rise, the extent of outmigration triggers other risks and insecurities. This raises the need for enhanced attention to adaptation to climate change.

Small island developing states (SIDS) face existential dangers from climate change. Their small land area, with low-lying coasts, isolation, and exposed position, and limited natural resources render them particularly vulnerable to sea-level rise, weather anomalies and storm surges. The IPCC Sixth Assessment Report (2022) projects severe climate impacts on SIDS in terms of tropical cyclones, coral reef depletion, saltwater intrusion into freshwater systems, and coastal erosion, which can lead to the loss of habitable land and livelihoods. In addition, the small institutional capacity and size of human resources, exacerbated by outmigration, limit states' ability to implement, monitor and evaluate actions that could help SIDS adapt to and cope under the conditions of climate change (Uitto et al. 2017). These challenges are not new in the context of climate change (Kelman 2013); but the current rapid changes accentuate them.

Countermeasures are called for to prevent small island states from losing both population and territory to the ravages of climate change. Nakayama et al. (2022) examined two main adaptive measures to cope with this situation: (a) migration to other countries; and (b) staying within their own countries. Each of these measures presented two possible options, leading to a scrutiny of four different alternatives: (i) migration to industrialized countries; (ii) migration to other island states; (iii) land reclamation and raising; and (iv) developing floating platforms. As experts, they rated these possible alternatives numerically and concluded that 'migration to industrialized countries' was the most favorable choice among the four alternatives. However, this alternative implies that an island state should lose both population and territory eventually, which is not acceptable for the countries themselves, nor for many in the international community concerned with sustainable development for people and national sovereignty. For island states to conserve people and land, Nakayama et al. (2022) propose the two possible alternatives of land reclamation and raising, and developing floating platforms, as suggested above. They analyze the relative advantages and disadvantages of these measures compared to migration and concluded that the alternative of 'land reclamation and raising' had the least adverse impacts (including mental unrest) on future climate migrants.

The Maldives, with the lowest average elevation among atoll countries, faces the risk of land loss due to rising sea levels. To counter this, the government is building Hulhumale, an artificial island, promoting in-country migration for better job opportunities. The decline in fishing and reluctance for physical work among young Maldivians contribute to urban migration. While Hulhumale attracts migrants with a high standard of living and education (Fig. 1), potential issues of loneliness are overlooked. To facilitate smooth migration, policies addressing occupational transitions, public job placement, and training programs are essential. Currently, mental health concerns are not prevalent, and sea level rise isn't a primary motivator for migration (Sakamoto et al. 2022).

SIDS receive significant attention and funding from international financial institutions focusing on environment and sustainable development. The Green Climate Fund (GCF), the largest dedicated climate finance fund, has as of June 2024 provided US$1.6 billion (12% of its total funding) to SIDS (GCF 2024). The Adaptation Fund provides funding for helping the most vulnerable countries, including SIDS, to cope with the impacts of climate change. The parties to the UN Framework Convention on Climate Change and the Paris Agreement agreed to establish a new funding arrangement in response to the losses and damages resulting from climate change. Philippines was selected in June 2024 as the host country for the Board of the new Loss and Damage Fund (UNFCCC 2024).

The Global Environment Facility (GEF) is the only one that funds projects and programs incorporating environmental issues beyond climate change. Its initiatives focus on maintaining and enhancing ecosystem goods and services through integrated approaches to managing land, water, forest, biodiversity, and coastal resources that also contribute to poverty reduction, sustainable livelihoods, and climate resilience

Fig. 1 Residential buildings in Hulhumalé, Republic of Maldives (Taken by Nakayama on June 5, 2022)

Fig. 2 Majuro Atoll, the capital of the Republic of the Marshall Islands (Taken by Nakayama on August 31, 2017)

(Batra and Norheim 2022). The 'Ridge-to-Reef' (R2R) initiative employs an integrated watershed management approach starting from the top of the island to the coral reef, which includes the prevention of flows of wastewater, chemicals, nutrients and sediments from land-based activities that degrade coastal ecosystems and resources. Land-use management, including deforestation control, is another key dimension of sustainability.

When outmigration is necessary, it needs to be managed appropriately. McClain, Bruch and Fujii (2022) point out that achieving 'migration with dignity,' which has become the keynote of migration policy today, requires several conditions, which are not necessarily met by current outmigration situations, such as that involving migration from the Republic of the Marshall Islands (RMI, Fig. 2) to the United States (U.S.) (O'Connor et al. 2019; Maekawa et al. 2019).

2 Data Sources and Case Studies

Drawing upon a set of studies, as well as evaluative evidence from on-the-ground initiatives, this chapter scrutinizes possible means to maintain peace and security in small island states of the Pacific and Indian Oceans, particularly atoll states, by enhancing their resilience to climate-induced risks, and by nurturing sustainable resource management policies and practices (for our study area, see Fig. 3).

A global SIDS evaluation was conducted by the GEF Independent Evaluation Office to assess the relevance and performance of GEF support aimed at addressing the main environmental challenges to SIDS from the country perspective; and to provide a deeper understanding of the determinants of sustainability (GEF IEO 2019). The evaluation took a mixed methods approach encompassing both quantitative and qualitative methods, covering 286 projects in the portfolio as well as country case studies conducted in 10 countries.

The chapter then contains a set of case studies conducted by the authors. Questionnaires were distributed at the University of the South Pacific (USP) Tarawa campus in Kiribati (Fig. 4) to explore students' perceptions of migration. Questionnaires were distributed to migrants from Kiribati in three major cities and towns in Fiji, namely Suva, Nadi, and Lautoka. Interviews were conducted with residents of Suva, Nadi and Lautoka. The research team also consulted with relevant authorities, researchers, international organizations and NGOs to understand the policy context (Maekawa et al. 2022).

The case study on migration from the RMI to the USA draws upon findings from a series of structured interviews, background interviews, and surveys with Marshallese living in Springdale, Arkansas. These interviews and surveys sought to understand the primary reasons for emigrating from the Marshall Islands, and gain perspective on their lives and livelihoods since arrival (McClain et al. 2020).

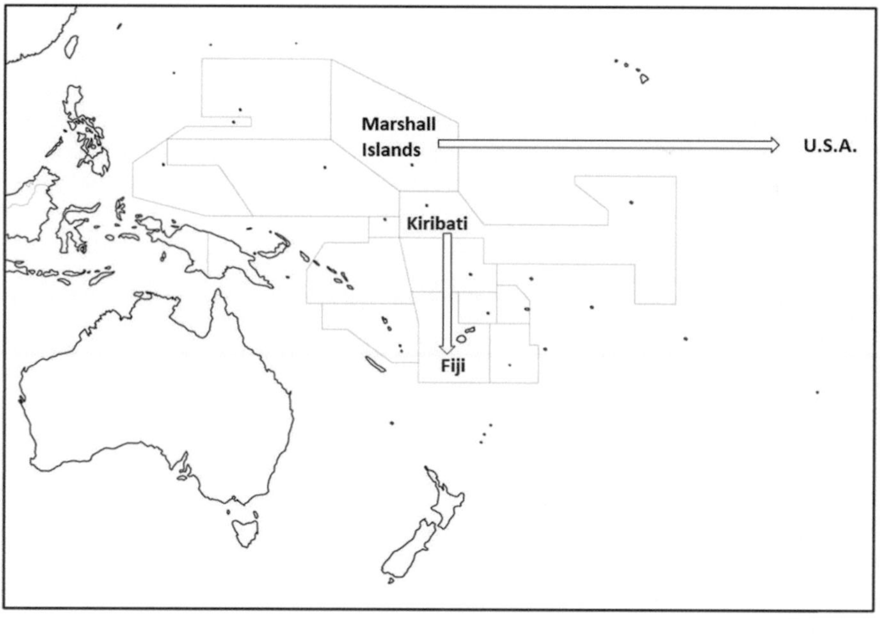

Fig. 3 Location of Pacific SIDS and the direction of main migration flows

Fig. 4 Tarawa Atoll, the capital of the Republic of Kiribati (Taken by Nakayama on February 27, 2019)

The chapter is also informed by a case study in the Maldives. Statistical data provided by the Maldivian government was analyzed to provide information on the changing lifestyle preferences of Maldivians. A questionnaire survey was conducted with residents of Hulhumale to determine the factors that explain changes in life satisfaction among those who have migrated to Hulhumale. Interviews were also conducted with individuals who had extensive knowledge of Maldivian society and Hulhumale. The interviewees were presented with a summary of the survey results and asked to interpret the results (Sakamoto et al. 2022).

3 Sustainable Development Challenges in SIDS

Small islands are at the forefront of the struggle for sustainable development. The United Nations 2030 Agenda for Sustainable Development and the Sustainable Development Goals (SDGs) linked thereto recognize the interlinkages between the social, economic and environmental dimensions of sustainability. Research has demonstrated the close interlinkages between the 17 SDGs and associated targets, as well as the inherent contradictions and necessary trade-offs between the goals (Issa, Mezher and El Fadel 2024; Laumann et al. 2022). While social, economic and environmental actions are all necessary for sustainable development, it is argued that biophysical and climatic sustainability provide the foundation on which human societies rest. Yet, too often the environmental dimensions of sustainable development are de-prioritized; and there is evidence that the environmental goals often receive secondary consideration in relation to the social and, especially, economic ones, and have a lower likelihood of being realized (Dawes, Zhou and Moinuddin 2022; Hickel 2019).

The United Nations Development Programme (UNDP 2024) has identified a set of specific characteristics that SIDS share and make them particularly susceptible to sustainability challenges: (i) remoteness, (ii) small populations, (iii) dependence on ocean resources, (iv) reliance on imports, (v) limited access to finance, (vi) debt pressure, and (vii) vulnerability to climate change. All of these factors render SIDS particularly vulnerable. The 4th International Conference on Small Island Developing States, organized by the UN in Antigua and Barbuda in May 2024, emphasized the value of the Multidimensional Vulnerability Index, which provides a comprehensive picture of a country's vulnerability across the core three domains of sustainable development and helps international financial institutions to incorporate appropriate actions into their policies and programs (UN 2024).

Climate change threatens the sustainable development of SIDS through multiple avenues, most notably extreme weather events, sea-level rise, high temperatures, and ocean acidification.

3.1 Integration of Nature-Based Solutions

SIDS in the Indian and Pacific Oceans face major challenges to sustainable development and environmental management, while possessing limited human resources and capacity to respond to these challenges. At the same time, they are 'large ocean states': custodians of vast marine areas with globally significant natural resources and biodiversity.

Established in 1991, the Global Environment Facility (GEF) is the most established public funding agency for environmental issues in developing states (GEF 2023). It also serves as an entity of the financial mechanism for five multilateral environmental agreements, including the UN Framework Convention on Climate Change, the Convention on Biodiversity, the UN Convention to Combat Desertification, and two conventions dealing with harmful chemicals. The GEF works with SIDS in the Pacific and Indian Oceans to assist them in maintaining and enhancing ecosystem goods and services through integrated approaches to land, water, forest, biodiversity, and coastal resource management, while contributing to poverty reduction, sustainable livelihoods, and climate resilience. Building resilience and capacity will in time result in reduced pressures on land and other natural resources and, consequently, dampen the inevitability for migration. The GEF work comprises both on-the-ground projects as well as mainstreaming environmental considerations into national policy and regulatory frameworks.

Almost a third of the GEF's SIDS projects have adopted integrated approaches to environmental and resource management. As mentioned above, R2R is an integrated watershed management program taking the whole island, from the top to the coral reef, as the target. Fourteen Pacific Island Countries (Cook Islands, Fiji, Kiribati, Marshall Islands, Micronesia, Nauru, Niue, Palau, Papua New Guinea, Samoa, Solomon Islands, Tonga, Tuvalu, and Vanuatu) participate in the Pacific R2R program. The latest GEF strategy launched the Blue and Green Islands Integrated Program, the objective of which is to apply nature-based solutions (NbS) in key ecosystems that support socio-economic development in SIDS, placing nature at the center of human wellbeing. While many GEF-funded projects have incorporated NbS elements for a long time, its systematic inclusion based on scientific principles is fairly new. A fundamental perspective sees NbS as delivering significant benefits to nature, as well as to humans, as opposed to seeing nature simply in service of social outcomes (Bierbaum, 2020). The key features of the program focus on integration, addressing "cross-cutting upstream challenges related to accounting and valuing of ecosystems, policy coherence, and domestic public and private sector resource mobilization, among other areas; and addressing landscape level challenges related to 3 key sectors for the SIDS context (tourism, food-fisheries/agriculture, urban development)" (GEF 2022: 62).

The evaluation by the GEF Independent Evaluation Office (GEF IEO, 2019) found that the GEF has achieved concrete environmental results in several areas, especially biodiversity conservation, reducing deforestation and land degradation, and improvements in water quality and availability. It has also contributed to socioeconomic outcomes, although the evidence is more limited. According to the evaluation, the GEF's greatest added value has been through institutional strengthening, capacity building, and assistance to legal and regulatory frameworks (Batra and Norheim 2022: 79–88). This is particularly important because of SIDS's limited institutional capacities.

Enhancing institutional capacity and the management of natural resources on a sustainable basis will contribute to resilience—both in terms of physical and social and economic responses towards climate change and other external shocks—and will also reduce migration pressures.

4 Migration and Development in SIDS

The limited human resources of SIDS pose a challenge to their ability to build resilient societies and economies and to cope with the various stresses on their sustainability. This is further exacerbated by outmigration. There are multiple reasons for migrating and the decision to leave one's country of origin is often not straightforward (Kelman 2015). Apart from limited livelihood opportunities at home and the decision to pursue education or healthcare overseas, climate change is increasingly a factor in the decision to migrate. According to UNDP (2024), SIDS lost US$153 billion due to extreme weather-related events between 1970 and 2020. This is a significant sum given that average GDP of SIDS is just US$13.7 billion. In short, sea level rise threatens the livability of small islands.

The International Organization for Migration (IOM 2024) highlights the complexity of migration pertaining to SIDS, which have long incorporated migration as part of their resilience strategy. Remittances from migrants provide larger financial resources to the countries of origin than official development assistance and direct foreign investment combined. These remittances are an important resource that contributes to both poverty reduction and adaptation to climate change. Yet, outmigration also depletes the human resources of the country. It is often the younger and better educated who migrate. According to IOM, SIDS are countries of origin of some 10 million international migrants. As noted below, a significant part of this movement involves intra-regional (inter-island) migration: SIDS host 3.8 million international migrants.

4.1 Outmigration as an Adaptation Strategy

Outmigration has emerged as a crucial adaptation strategy (McLeman and Smit 2006; Gemenne and Blocher 2017) for people from island countries facing the escalating threats associated with climate change. Many residents of these countries are compelled to consider or seek safer and more stable environments abroad (Barnett 2001). Outmigration offers both an escape from immediate environmental and public health risks as well as provides longer-term potential for better economic and educational opportunities, access to resources, and improved living conditions in host countries.

It should be noted that, as mentioned above, 'migration to industrialized countries,' which may be appealing at some level, may lead to substantial costs for migrants seeking a better livelihood abroad. Such costs include the mental pressure caused by leaving one's own language and culture—with which one would most likely have been familiar since birth—to live in a foreign country with a different language and culture. Younger generations may be more tolerant of such changes, while for older people such changes are likely to be an insurmountable barrier. Furthermore, cultural preservation has intrinsic value and is an important contributor to sustainable development (UNESCO 2023).

The decision to migrate as an adaptation strategy is complex and multifaceted (Kelman 2015). Climate migrants from island states face significant challenges in making the difficult choice of leaving their homes and communities behind (Fujikura et al. 2023). Often, the decision is driven not only by the current climate-related threats but also by the anticipation of future climate impacts, which can exacerbate the vulnerabilities of their home countries over time, as well as by more immediate and even potentially temporary concerns driving migration (such as education, jobs, and healthcare).

As outmigration continues, it inevitably leads to changes in the demographic structure of island states, affecting their social fabric, culture, and economic dynamics.

For instance, in the case of the RMI, the population declined from 53,158 to 39,262 between 2011 and 2021. Additionally, the loss of skilled and productive individuals (so-called brain drain) can further strain the resilience and adaptive capacity of island communities already grappling with climate-induced disruptions (Maekawa et al. 2022).

While outmigration presents a viable adaptation option for some climate migrants, it also raises ethical, legal, and policy concerns. The migration process can be arduous and perilous, subjecting migrants to risks such as exploitation, human trafficking, and human rights violations. Such difficulties can cause psychological problems for some migrants, and measures are needed to deal with the problems they face (Fujikura et al. 2019; Surjan et al. 2016).

Moreover, the burden of climate-induced migration often falls disproportionately on vulnerable populations, including the poor, marginalized groups, and Indigenous communities, who may have limited resources to navigate the migration process.

Addressing these challenges requires a comprehensive approach that involves cooperation between source and destination countries, international organizations, and civil society to ensure the protection and wellbeing of climate migrants. Effective policy frameworks are required to facilitate safe, orderly, and dignified migration, while simultaneously working towards mitigating and adapting to climate change at home so as to reduce the need for forced displacement in the first place.

4.2 International Migration Case Studies

This section presents two case studies on international migration: one from Kiribati to Fiji, and the other from the Marshall Islands to the United States. Kiribati and the Marshall Islands are both atoll countries severely threatened by the impacts of climate change. Atoll countries are composed of coral reefs, and their average elevation is as low as 1 to 2 m above sea level. Therefore, among all countries, atoll countries are most affected by climate change-driven sea level rise.

Case 1: Migration from Kiribati to Fiji

Atoll countries need to pursue outgoing alternatives to navigate the challenges ahead. The destinations to which the nationals of atoll countries migrate are, in reality, limited to some developed countries and other island states in hegemonic positions in terms of regional dominance, such as Fiji which has the largest economy, population and landmass in the Pacific and is host to regional institutions like the University of the South Pacific (Maekawa et al. 2019).[1]

The impacts of climate change are strongly felt in the Pacific islands, especially in Kiribati. The Republic of Kiribati is located in the central Pacific Ocean and consists of low-lying atolls with an average elevation of less than two meters with 811 km^2 of land area, with an estimated population of 113,140 in 2017. The population of Fiji

[1] Not including Australia, New Zealand and Papua New Guinea.

was estimated at 884,887 (Fiji Bureau of Statistics 2023), and it has a land area of 18,270 km^2 (Country Information 2023).

The citizens of Kiribati (known as I-Kiribati) have been anxiously observing the impacts of climate change in the forms of coastal erosion, king tides, and intensifying cyclones, among others. Maekawa et al. (2019) report that Kiribati nationals often express a desire to migrate abroad, particularly after completing higher education, in pursuit of better education, employment opportunities, and partially influenced by the effects of climate change. The study investigated the real-life experiences of Kiribati nationals in establishing themselves and their livelihoods in Fiji.

Fiji has better educational and medical facilities than other Pacific Island states. However, the scale of economic activity is not sufficient to attract large numbers of immigrants from other island states to settle and work in Fiji. In terms of education, there are only a limited number of fields of study in which advanced research can be conducted at graduate schools in Fiji, and there are only a limited number of specialized fields in which international students from other island states can receive a satisfactory level of education or be mentored in research. Because of these existing constraints, migration to other island states is likely to be limited, and the main destinations for outmigration will be developed countries.

Case 2: *The Case of Marshallese Migrants in the United States*

Marshallese immigrants in Springdale, Arkansas, United States, have been a growing community in recent years. The Marshallese population in Springdale has been attracted by employment opportunities in the local poultry industry and other sectors, leading to a significant increase in their numbers in the region (Duke 2014). As a result, Springdale has become one of the largest Marshallese communities in the continental United States.

The Compact of Free Association, a bilateral agreement between the RMI and the U.S., grants Marshallese citizens the opportunity to live and work in the U.S. indefinitely, free of the usual visa and quota requirements (U.S. Department of State 2003). Under this agreement, they can freely travel between the RMI and the U.S., benefiting from unrestricted mobility between the two countries.

The Marshallese community in Arkansas began development during the 1980s, and over the period between 2000 and 2010, the population experienced a notable surge (Jimeno and Rafael 2013). Presently, Springdale, Arkansas, is home to an estimated 17,000 Marshallese residents, and this number continues to rise not just in Springdale but also in other parts of the USA, such as the states of California, Hawai'i, New Mexico, Oklahoma, Oregon, and Washington. Many individuals relocated to the area in pursuit of better job prospects, improved access to education and healthcare, and for family reasons (McClain et al. 2020). However, adjusting to life in Springdale has presented numerous challenges for a significant portion of the Marshallese population.

Migration to the U.S. from RMI remains a prevalent trend, with one-third of RMI's population having migrated to the U.S., and this number continuing to grow annually. Marshallese individuals residing in Springdale often report that family and healthcare were compelling reasons for their move (Fujikura et al. 2023).

Many immigrants were invited by family members already living in Springdale to seek medical care, assist in child-rearing, or pursue education. The decision to move to the U.S. was often sudden for many, with 42% indicating that they moved without much advance notice, often at the behest of family or friends already living in the US, 30% had 1–3 months to plan their move, and 21% took more than 3 months, with none exceeding a year in planning their migration (McClain et al. 2020). They arrived in Springdale with little understanding of American culture and customs.

The lack of awareness regarding the significance of English in the U.S. has led to a considerable number of Marshallese immigrants coming to the country without a proper understanding of the language's importance. Consequently, just 29% of Marshallese immigrants in Springdale consider themselves fluent in English (McClain et al. 2020). Limited proficiency in English presents significant challenges. Those engaged in low-skilled jobs, such as working in a poultry processing facility, can manage to survive within the Marshallese community, relying on familiar customs and language. However, to pursue higher education and better job prospects, fluency in English is crucial. Although local NGOs, schools and businesses offer English language courses, only a small fraction of the Marshallese community takes advantage of these opportunities (Fujikura et al. 2019). Reasons for low uptake include family commitments and transportation challenges.

Social ties among immigrants represent the relationships they form with each other. Typically, these bonds serve as valuable sources of emotional and financial support for those who have migrated abroad. For the Marshallese community in the United States, however, these social ties sometimes have counterproductive effects. Within their community and culture, the term "family" encompasses a vast extended network. Older men hold dominant positions within these families, and younger members are expected to abide by the decisions of their elders. According to Marshallese immigrants in Springdale, this cultural dynamic persists among immigrants in the U.S., often resulting in the marginalization of young immigrants or second-generation Marshallese (Fujikura et al. 2019).

Conversely, there is a lack of strong social connections between the Marshallese community and the host community in Springdale. The Marshallese people's social shyness and language barriers further contribute to their detachment from the host community. While the host community is generally welcoming, the scarcity of interactions hinders the development of robust bridging social capital between them (Fujikura et al. 2019).

4.3 Comparative Analysis of Atoll State Migration: Marshallese and I-Kiribati Experiences

The cases of Marshallese migration to the United States and I-Kiribati migration to Fiji offer insights into the diverse challenges and opportunities faced by migrants from atoll states. While these cases present different contexts, a comparative analysis reveals important commonalities and distinctions that can inform our understanding of climate-induced migration from small island developing states.

Similarities:

Push Factors: Both Marshallese and I-Kiribati migrants are driven by similar push factors, including limited economic opportunities at home, the desire for better education and healthcare, and the looming threat of climate change impacts on their low-lying islands.

Education as a Key Driver: The pursuit of better educational opportunities emerges as a primary motivation for migration in both cases, highlighting the importance of human capital development in migration decisions.

Cultural Challenges: In both cases, migrants face challenges in adapting to new cultural contexts, particularly in terms of language barriers and social integration.

Reliance on Social Networks: Both Marshallese and I-Kiribati migrants heavily rely on family and community connections in their destination countries for support during the migration and settlement process.

Differences:

Legal Frameworks: The Compact of Free Association between the Marshall Islands and the United States provides a particular legal pathway for Marshallese migration, which is not available to I-Kiribati migrants to Fiji. This difference significantly impacts the ease of migration and the rights afforded to migrants in their destination countries.

Scale of Migration: The scale of Marshallese migration to the United States is much larger than I-Kiribati migration to Fiji, partly due to the differences in legal frameworks and the economic capacities of the destination countries.

Cultural Proximity: I-Kiribati migrants to Fiji benefit from greater cultural proximity and similarity in Pacific Island contexts, potentially easing some aspects of cultural adaptation compared to Marshallese migrants in the United States.

Access to Education and Health: Access to education and health forms an important aspect of Migration with Dignity. Especially, access to health in Fiji and U.S. presents a distinct contrast, where in Fiji, all migrants have the same access as citizens to state-funded services, whereas in the U.S., access to health services is more challenging.

Economic Integration: The economic integration experiences differ, with Marshallese migrants in the U.S. often finding employment in specific sectors (e.g., poultry processing), while I-Kiribati in Fiji face a more diverse but potentially more limited job market.

By examining these cases side by side, we gain a more nuanced understanding of the migration landscape for atoll states. While the approaches and contexts differ, they illustrate many of the diverse challenges and opportunities in climate-induced migration.

4.4 Migration with Dignity

Recognizing that climate change has been and will continue to be a major driver of migration, the international community has struggled to develop policies to support and govern migration. For years, scholars and advocates have called for a convention on climate refugees (Berchin et al 2017; Ferracioli 2014; Bierman and Boas 2008); but political and technical challenges have thwarted agreement on such an instrument. The 2018 Global Compact for Safe, Orderly and Regular Migration provides a non-binding suite of measures covering a wide range of aspects of migration (UNGA 2018). This is an important start, but its non-binding nature means that it does not actually require government action; nor does it provide enforceable protections (O'Connor et al. 2019).

Responding to these needs, the Migration with Dignity Framework provides a conceptual and operational framework that draws upon existing "dignity rights," already found in international human rights instruments and across national constitutions (McClain et al. 2022). It emphasizes six core clusters of rights: freedom of movement, right to be secure, right of equality, rights to a basic quality of life, right to access services, and civil and political rights (see Table 1).

The 'Migration with Dignity' Framework is important for three key reasons. First, since it draws upon international human rights law and widespread national constitutional law, most of the elements of the Framework are already in force; and so these rights are already largely recognized (McClain et al. 2022). Indeed, and arguably, the only right that is not well-developed in existing international and constitutional law is the right to enter a country (McClain et al. 2020). Second, because of its roots in human rights and constitutional law, the framework applies to a broad range of contexts, from people fleeing conflict and persecution (i.e., refugees), to people traveling for work (migrant labor), to those seeking a better life (economic migrants), to people fleeing climate change and environmental degradation. Reasons for migration are often multiple and complicated (O'Connor et al. 2019). Because of its breadth of relevance, the actual reasons for migration are secondary. Third, the Framework extends considerations of migration well past the issue of immigration (movement) to considering essential the broader dimensions of a life with dignity.

In the context of climate change, the 'Migration with Dignity' Framework provides a much-needed framework based on existing law. While a new convention on climate migrants seems unlikely, the Migration with Dignity Framework provides an enabling legal framework that is based on existing law (particularly international human rights law and comparative constitutional law). Moreover, it side-steps both questions of persecution (necessary for someone to be declared a refugee) as well as those of motivation. Finally, it addresses important aspects of life after immigration, including the ability to find a job and access services (McClain et al. 2022).

This section draws on a case study conducted over 2018–2019 to examine overall livelihood situations, perceptions, and motivations of I-Kiribati migrating to Fiji, covering 91 respondents in Kiribati who have a potential to migrate and 77 respondents who had actually migrated to Fiji from Kiribati (Maekawa et al. 2019).

Table 1 The migration with dignity framework. *Source* Adapted from McClain et al. (2022)

Freedom of movement	Right to be secure	Right of equality	Rights to a basic quality of life	Right to access services	Civil and political rights
Freedom to leave country of origin	Free from sexual violence (including rape and sexual exploitation)	Non-discrimination and equal treatment under the law (based on race, color, gender, language, religion, political, national origin, birth, or status)	Employment rights	Health care	Identity
Freedom to return/re-enter country of origin	Free from human trafficking, slavery, and forced labor		Housing rights	Education	Religion
Freedom of movement with country of origin or country of settlement	Free from arbitrary and abusive detention	Opportunities for upward mobility	Food rights	Utilities	Language
Admission to a foreign country freedom to stay				Legal services	Free speech

According to this study, Fiji was the most popular destination for cross-border migration for Kiribati population, followed by Australia and New Zealand. The main motivations for migration are receiving a better education, and getting a job, in that order. Sea-level rise is also indicated as a significant reason for migration, followed by others such as family and health.

Table 2 indicates the available legal and policy measures, especially international conventions for migrants who cross borders. In fact, Fiji has signed and ratified most of the international conventions related to migration.

In order to examine the implementation of these policies and to understand the actual livelihoods situations of the migrants from Kiribati to Fiji from a 'Migration with Dignity' perspective, a survey and follow-up interviews were conducted as mentioned above. I-Kiribati tend to have a desire to migrate abroad, especially after obtaining higher education, for further education and good jobs, and to some extent due to impacts of climate change.

People's motivations and aspirations typically remain consistent before and after relocating, and a majority of individuals find that life in Fiji surpasses their initial expectations. However, they encounter challenges when it comes to improving their

Table 2 Signature and ratification of international conventions by Fiji

Convention name	Ratified (Year)
International Labour Organization (ILO) migration for employment convention (Revised), 1949 (No. 97)	No
United nations convention relating to the status of refugees (also known as the refugee convention), 1951	Yes (1972)
United Nations conventions on statelessness, 1954 and 1961	Yes (1972, 1966)
ILO migrant workers (Supplementary Provisions) convention, 1975 (No. 143)	No
United Nations Convention on the Rights of the Child (UNCRC), 1989	Yes (1993)
United nations international convention on the protection of the rights of all migrant workers and members of their families (ICRMW), 1990	Yes (2019)

Source Table 1, IOM (2022: 12)

education and language skills, especially in English. When searching for employment, English proficiency and professional qualifications secured abroad (e.g., medical, legal, construction, automotive) are often not acknowledged in Fiji, leading many Kiribati residents in Fiji to enroll in schools to obtain the necessary qualifications for employment opportunities in the country. There seem to be no typical jobs that I-Kiribati engage in. The respondents who already had job experience and qualifications (especially those acquired from such places as Australia and New Zealand) had a relative advantage in finding similar jobs in Fiji smoothly. The ones who did not have qualifications to be recognized in Fiji tried to find ways to gain training and education in Fiji if they could afford them. Those who have adequately prepared in advance tend to secure jobs smoothly following their relocation. When establishing themselves in their new environment, individuals tend to rely on family connections, community groups, and similar resources as ongoing sources of support, security and finance.

In terms of access to healthcare services, all migrants in Fiji have the same access as citizens to state-funded health services. As reported by IOM (2022), migrants to Fiji can access public healthcare, free of charge without identification. This is based on the Bill of Rights in the Constitution of the Republic of Fiji (2013), which promotes "the right of every person to health, and to the conditions and facilities necessary to good health, and to health care services, including reproductive health care" (IOM 2022: 12). Fiji is also a signatory of the Astana Declaration (2018) committing the country to the "fundamental right of every human being to the enjoyment of the highest attainable standard of health without distinction of any kind" (ibid.). The survey respondents confirmed that they enjoyed access to healthcare similar to other Fiji residents.

Overall, 19% of I-Kiribati migrants in Fiji who responded to our survey indicated climate change as a factor for their migration, but their primary reasons were to seek better education and employment. Migrants from Kiribati enjoy access to public education and healthcare, although job-seeking is left to individuals' efforts. When seeking employment, their lack of English-language skills and professional qualifications posed challenges. Those who had prepared well in advance tended to find jobs relatively smoothly after migration. This sort of understanding and statistics on migrants is important in informing migration policies of countries of origin, transit, and destination.

While 'Migration with Dignity' has generally been applied to contexts where people have migrated, it necessarily includes the right to stay (sometimes referred to as the "right to remain"). For Anote Tong, former prime minister of Kiribati who coined the term, 'Migration with Dignity' enables people to move if and when they want to move (Tong 2014). It does not require or force people to move.

The scope of the right to stay includes at least five rights: the freedom to choose one's residence, the right to not have one's property taken (except for public purpose and, if so, with just compensation), the right to live a life with dignity, the freedom of movement, and the rights of identity and culture (often tied to a particular geography). These rights are enshrined in international conventions, including the International Covenant on Civil and Political Rights (arts. 10, 12, 27) and the Universal Declaration of Human Rights (arts. 1, 13, 17, 22, 27). Regional agreements are also important: the American Convention on Human Rights (arts. 11, 21, 22); the African Charter on Human and Peoples' Rights (arts. 5, 12, 14, 22, 24); and the European Convention on Human Rights (art. 2; Protocol, art. 1). National constitutions frequently protect one or more of these rights. In addition, the principle of legal certainty recognizes the right of individuals to make long-term plans for their lives by requiring that state action be reasonably predictable and nonarbitrary.

The right to stay is a qualified right, subject to restrictions provided by law to protect national security, public order, public health, public morals, and the rights and freedoms of others. The right to stay on a particular property is often subject to payment of property taxes. In most countries, land may be expropriated for a public purpose and for which just compensation is provided. Moreover, there are distinctions between nationals and aliens regarding the right to stay, as aliens may not be able to own property (in some countries), and countries can deport aliens pursuant to a decision reached in accordance with due process.

There are some outstanding questions regarding the right to stay. For example, can a person stay with their family if it places children at risk? The answer is uncertain and likely depends both on the jurisdiction and on how imminent the harm is. Second, does a state have an obligation to continue providing services to a particular property? That is to say, does a state have to provide roads, electricity, and other services to someone who chooses to stay? A decision to no longer provide these services is not generally considered as requiring compensation (at least based on global jurisprudence to date). Does a right to stay imply a right to conditions that enable people to lead a good life in their homeland (Hoesch 2022; Oberman 2011)? International law does not yet require this, but that may change, particularly considering the emergence of 'Loss and Damage' claims under the UN Framework Convention on Climate Change. Similarly, international law does not yet ensure that people are able to stay on land that is rapidly increasing in value (i.e., gentrifying) such that they may not be able to afford to pay the property taxes necessary to stay on the land (Newman and Wyly 2006).

In summary, there is a right to stay under international human rights law. There are boundaries and limitations on the right to stay. And there are outstanding questions regarding the implications of the right to stay in a changing climate.

5 Conclusion

This chapter has discussed the implications of outmigration from island states, possible outmigration alternatives for atoll states, and highlighted two cases from such atoll states: Marshallese migrants in the United States, and Kiribati migrants to Fiji. These findings highlight four key conclusions. First, outmigration is a complex and multifaceted phenomenon, with both positive and negative consequences for island states and their citizens. Second, the main destinations for outmigration from atoll states are likely to be larger, more developed states, as other island states lack the capacity to accommodate large numbers of immigrants. Third, migrants face a number of challenges, including language barriers, limited social connections with the host community, and cultural dissonance. Fourth, there is a need for more support services for immigrants, such as English language classes, cultural orientation programs, and help with job placement.

While migration remains an adaptation option for many islanders, it would seem preferable to find solutions that would reduce the need for leaving the country of origin. Organizations, such as the GEF, are supporting sustainable development in SIDS to strengthen their resilience towards climate and other external shocks. These solutions must take an integrated approach that addresses both management of natural resources (water, land, coastal zone, fisheries, etc.) as well as brings social and economic benefits to the population. Strengthening local institutions forms an important dimension.

While international migration can serve as an adaptation strategy for individuals facing climate change impacts in SIDS, it is crucial to consider the potential negative consequences for the countries of origin. Promoting and encouraging outmigration may lead to population decline and economic challenges in island nations, potentially undermining their long-term sustainable development prospects.

Population decline leads to a loss of skilled workers. This 'brain drain' effect can result in reduced economic productivity, and weakened social structures, which are essential for maintaining resilient communities. From a sustainable development perspective, it is imperative to strike a balance between supporting individual adaptive strategies and preserving the viability of SIDS. This underscores the importance of implementing comprehensive in situ adaptation measures, as discussed earlier, to enhance local resilience and create opportunities that encourage skilled individuals to remain in or return to their home countries. Additionally, policies that facilitate circular migration or knowledge transfer from diaspora communities could help mitigate the negative impacts of outmigration while still allowing individuals to benefit from international opportunities.

While nature-based solutions, such as protecting coastal mangroves and other vegetation, can be powerful, there may be a need for more structural strategies elsewhere, in particular in low-lying atoll states. These include infrastructure works for land reclamation, as well as the development of artificial islands, such as in the case of the Maldives. SIDS governments should develop comprehensive adaptation plans that explore a variety of options. These plans should take into account natural environmental factors (including nature-based solutions) as well as human and institutional aspects. They have to consider migration realistically. International organizations and wealthy countries should support these processes through technical and financial means. It is also important to facilitate safe and orderly migration for climate migrants.

These recommendations are some important steps towards supporting sustainable development and resilience in situ, as well as ensuring the protection and well-being of climate migrants from island states.

References

Barnett J (2001) Adapting to climate change in Pacific Island countries: the problem of uncertainty. World Dev 29(6):977–993

Batra G, Norheim T (2022) Staying small and beautiful: enhancing sustainability in the small island developing states. In: Uitto JI, Batra G (eds) Transformational change for people and the planet: evaluating environment and development, pp 73–91. Cham, Switzerland: Springer. https://link.springer.com/chapter/. https://doi.org/10.1007/978-3-030-78853-7_6. Accessed 17 Jan 2024

Berchin II, Valduga IB, Garcia J, de Andrade Guerra JB (2017) Climate change and forced migrations: an effort towards recognizing climate refugees. Geoforum 84:147–150

Bierbaum R (2020) Nature-based solutions and the GEF. In: A STAP advisory document. Washington, DC: Scientific and Technical Advisory Panel to the Global Environment Facility. https://stapgef.org/resources/advisory-documents/nature-based-solutions-and-gef. Accessed 17 Jan 2024

Bierman F, Boas I (2008) Protecting climate refugees: the case for a global protocol. Environment 50(6):8–17

Dawes JHP, Zhou X, Moinuddin M (2022) System-level consequences of synergies and trade-offs between SDGs: quantitative analysis of interlinkage networks at country level. Sustain Sci 17:1435–1457

Duke MR (2014) Marshall islanders: migration patterns and health-care challenges. Online J Migr Policy Inst. https://www.migrationpolicy.org/article/marshall-islanders-migration-patterns-and-health-care-challenges. Accessed 17 Jan 2024

Ferracioli L (2014) The appeal and danger of a new refugee convention. Soc Theory Pract 40(1):123–144

Fiji Bureau of Statistics (2023) 2017 census results. https://www.fiji.gov.fj/Media-Centre/News/Fiji-Bureau-of-Statistics-Releases-2017-Census-Res. Accessed 17 Jan 2024

Fujikura R, Nakayama M, Sasaki D, Taafaki I, Chen J (2023) Family and community obligations motivate people to immigrate: a case study from the Republic of the Marshall Islands. Int J Environ Res Public Health 20(8):5448

Fujikura R, Nakayama M, McClain SN, Drinkall S (2019) Addressing the health problems after immigration faced by the Marshallese in Springdale, Arkansas: lessons learned from the city of Vienna. J Disaster Res 14(9):1309–1316. https://www.fujipress.jp/jdr/dr/dsstr001400091309/. Accessed 17 Jan 2024

GCF (2024) GCF spotlight: small island developing states. In: Green climate fund. https://www.greenclimate.fund/document/gcf-small-island-developing-states-factsheet. Accessed 18 Sept 2024

GEF IEO (2019) Strategic country cluster evaluation: small island developing states. Washington, DC: Independent Evaluation Office of the Global Environment Facility. https://www.gefieo.org/evaluations/scce-sids. Accessed 17 Jan 2024

GEF (2022) GEF-8 programming directions. GEF/R.08/17. Washington, DC: Global Environment Facility and World Bank Group. https://www.thegef.org/sites/default/files/documents/2022-01/GEF_R.08_17_GEF-8_Programming_Directions.pdf. Accessed 17 Jan 2024

GEF (2023) Global environment facility. https://www.thegef.org/. Accessed 17 Jan 2024

Gemenne F, Blocher J (2017) How can migration serve adaptation to climate change? Challenges to fleshing out a policy ideal. Geogr J 183(4):336–347

Hickel J (2019) The contradiction of the sustainable development goals: growth versus ecology on a finite planet. Sustain Dev 27:873–884

Hoesch M (2022) A moral right to stay. Universität Münster. https://www.uni-muenster.de/news/view.php?cmdid=12673. Accessed 17 Jan 2024

Country Information (2023) Fiji Islands. https://pic.or.jp/country_information/4810/. Accessed 17 Jan 2024

IOM (International Organization for Migration) (2022) Migration governance indicators profile 2022—Republic of Fiji. Geneva: IOM

IOM (2024) IOM and the small island developing states. International Organization for Migration. https://migration4development.org/sites/default/files/2024-07/IOM%20and%20the%20SIDS_Factsheet%202024_FINAL.pdf. Accessed 5 Sept 2024

IPCC (Intergovernmental Panel on Climate Change) (2022) Climate change 2022: impact, adaptation and vulnerability. In: Sixth assessment report. Geneva: Switzerland. https://www.ipcc.ch/report/ar6/wg2/. Accessed 17 Jan 2024

Issa L, Mezher T, El Fadel M (2024) Can network analysis ascertain interlinkages towards evidence-based policy planning? A systematic critical assessment. Environ Impact Assess Rev 102:107295. https://doi.org/10.1016/j.eiar.2023.107295

Jimeno S, Rafael A (2013) A profile of the Marshallese community in Arkansas, vol 3. Little Rock: Winthrop Rockefeller Foundation and Migration Policy Institute

Kelman I (2013) No change from climate change: vulnerability and small island developing states. Geogr J 180(2):120–129

Kelman I (2015) Difficult decisions: migration from small island developing states under climate change. Earth's Futur 3(4):133–142. https://agupubs.onlinelibrary.wiley.com/. https://doi.org/10.1002/2014EF000278. Accessed 17 Jan 2024

Laumann F, von Kügelgen J, Kanashiro Uehara TH, Barahona M (2022) Complex interlinkages, key objectives, and nexuses among the sustainable development goals and climate change: a network analysis. Lancet Planet Health 6:e422-430

Maekawa M, Singh P, Charan D, Yoshioka N, Uakeia T (2019) Livelihood re-establishment of emigrants from Kiribati in Fiji. J Disaster Res 14(9):1277–1286. https://www.fujipress.jp/jdr/dr/dsstr001400091277/. Accessed 17 Jan 2024

Maekawa M, Nakayama M, Fujikura R, Yoshida T, Shiiba N (2022) Highly skilled migrant workers as a vulnerability of small island developing states during the COVID-19 pandemic: cases of three countries in Micronesia. J Disaster Res 17(3):380–387. https://www.fujipress.jp/jdr/dr/dsstr001700030380/. Accessed 17 Jan 2024

McClain SN, Bruch C, Nakayama M, Laelan M (2020) Migration with dignity: a case study on the livelihood transition of Marshallese to Springdale, Arkansas. J Int Migr Integr 21:847–859

McClain SN, Bruch C, Fujii M (2022) The methodology and application of a migration with dignity framework. J Disaster Res 17(3):308–314. https://www.fujipress.jp/jdr/dr/dsstr001700030308/. Accessed 17 Jan 2024

McLeman R, Smit B (2006) Migration as an adaptation to climate change. Clim Change 76(1–2):31–53

Nakayama M, Fujikura R, Okuda R, Fujii M, Takashima R, Murakawa T, Sakai E, Iwama H (2022) Alternatives for the Marshall islands to cope with the anticipated sea level rise by climate change. J Disaster Res 17(3):315–326. https://www.fujipress.jp/jdr/dr/dsstr001700030315/. Accessed 17 Jan 2024

Newman K, Wyly EK (2006) The right to stay put, revisited: gentrification and resistance to displacement in New York City. Geogr Urban Stud Plan 43(1):23–57

O'Connor S, Bruch C, Maekawa M (2019) Legal and policy options for environmental migrants. J Disaster Res 14(9):1254–1261. https://www.fujipress.jp/jdr/dr/dsstr001400091254/. Accessed 17 Jan 2024

Oberman K (2011) Immigration, global poverty and the right to stay. Polit Stud 59(2):253–268

Sakamoto A, Nishiya K, Guo X, Sugimoto A, Nagasaki W, Doi K (2022) Mitigating impacts of climate change induced sea level rise by infrastructure development: case of the Maldives. J Disaster Res 17(3):327–334. https://www.fujipress.jp/jdr/dr/dsstr001700030327/. Accessed 17 Jan 2024

Surjan A, Kudo S, Uitto JI (2016) Risk and vulnerability. In: Uitto JI, Shaw R (eds) Sustainable development and disaster risk reduction. Springer, Tokyo, pp 37–55

Tong A (2014) Statement by H.E. President Anote Tong at the 69th UN General Assembly. September 26. https://www.un.org/en/ga/69/meetings/gadebate/pdf/KI_en.pdf. Accessed 17 Jan 2024

Uitto JI, Kohlitz J, Todd D (2017) Evaluating sustainable development in the SIDS—lessons from the Pacific and the Caribbean. In: van den Berg RD, Naidoo I, Tamondong SD (eds) Evaluation for the agenda 2030: providing evidence on progress and sustainability, pp 119–133. Exeter, UK: IDEAS. https://ideas-global.org/wp-content/uploads//2017/12/Chapter-8.pdf. Accessed 17 Jan 2024

UN (2024) Report of the fourth international conference on small island developing states. Antigua and Barbua, 27–30 May 2024. New York: United Nations. https://sdgs.un.org/conferences/sids2024. Accessed 5 Sept 2024

UNDP (2024) Small island developing states are on the frontlines of climate change—here's why. United Nations Development Programme. https://climatepromise.undp.org/news-and-stories/small-island-developing-states-are-frontlines-climate-change-heres-why. Accessed 5 Sept 2024

UNESCO (2023) Culture for sustainable development. https://en.unesco.org and https://www.unesco.org/en/sustainable-development/cultureelopment/culture. Accessed 17 Jan 2024

UNFCCC (2024) Fund for responding to loss and damage. In: UN framework convention on climate change. https://unfccc.int/loss-and-damage-fund-joint-interim-secretariat. Accessed 18 Sept 2024

UNGA (2018) Global compact for safe, orderly and regular migration. A/RES/73/195. United Nations General Assembly. Adopted 19 December. https://www.un.org/en/development/desa/population/migration/generalassembly/docs/globalcompact/A_RES_73_195.pdf. Accessed 17 Jan 2024

US Department of State (2003) Compact of free association, 2003. https://2009-2017.state.gov/documents/organization/173999.pdf. Accessed 17 Jan 2024

'Illegal' Korean Immigrants in Island(ed) Detention Camps and Enduring Colonial Sovereignty: The Case of Sasebo, Japan, Under U.S. Occupation

Taku Fukumoto

Abstract This chapter focuses on the persistence of colonial elements in national sovereignty through the case of the internment camps for "illegal" Korean immigrants in Sasebo, Japan. The islands in the strait area between Japan and the Korean Peninsula, under the U.S. occupation, served as a stage where many irregular migrants crossed from Korea to Japan. Through their suppression, the national borders were re-demarcated. Historical documents show that the practices in the "spaces of exception" in Sasebo materialized the dialectic of exclusion and inclusion, in which the arbitrary "illegitimacy" was constructed, and all the former colonial people were subjected to authorities' extralegal demands. Sasebo, as the hinterland of the islands in this strait, also exemplified the fragmented nature of the marginality and centrality of these islands.

Keywords Detention camp · Spaces of exception · Colonialism · U.S. occupation of Japan · Sasebo

1 Introduction

Numerous efforts have been made to investigate the relationship between irregular migrants crossing national borders and their accommodation in detention facilities. Studies in this area have shed light on the treatment of migrants in detention facilities, thus enhancing our understanding of contemporary forms of sovereignty, border control, and identification/ordering of people. At the same time, some studies have pointed out the relationship between the geopolitical aspects of islands and the control of these migrants. This chapter analyzes the concrete processes of colonial power's

This chapter is a substantially revised version of Fukumoto (2013), written in Japanese, based on subsequent research developments.

T. Fukumoto (✉)
Department of Japanese Studies, Nanzan University, Nagoya, Japan
e-mail: fukutaku@nanzan-u.ac.jp

T. Yamazaki and G. Baldacchino (eds.), *Relational Geographies of Islands in the Indo-Pacific*, International Perspectives in Geography 23,
https://doi.org/10.1007/978-981-96-7155-7_6

survival in parallel with demarcational practices around islands and their ancillary areas, focusing on the example of "illegal" Korean entrants to Japan in the detention camp of Sasebo, Nagasaki Prefecture, during the U.S. occupation of Japan (1945–1952).

With reference to Giorgio Agamben's notion of "states of exception" (Agamben 2005), attempts have been made to clarify the (undemocratic) nature of state sovereignty and the classification of ruling/ruled in the modern world through the various practices that occur in detention centers (Mountz et al. 2012). The numerous cases of detentions have been framed under such terms as colonialism or post-colonialism. These terms seem appropriate to refer to the practice of sovereignty in a colonial manner; extralegal suppression outside suzerains' ordinary law systems, the creation of social categories of individuals to be detained, the identification of colonized people, the control over borders and mobility, among others.

"Spaces of exception" where the state of exception manifests itself, have attracted the particular attention. In the contemporary world, "spaces of exception" tend to be located and operationalized in the peripheral areas of nation-states, including islands. Some of these areas are sometimes outside the jurisdictions of regular legal systems, and certain locations have been designated as military bases. Detention camps where severe treatments toward irregular migrants occur are typically located on islands or in peripheral areas, located far from the mainlands of their parent state, such as the U.S.-administered Guantanamo Bay facility in Cuba, the "export" of the European Union's national borders, the U.K.'s evolving Rwanda detention policy, and Australia's offshore policy. As Mountz (2011) explains, these detention camps form "the enforcement archipelago."

Her expression is a kind of metaphor. At the same time, however, we must also pay attention to the historical geographies of respective detention camps. For example, the imprisonment of irregular migrants in Guantanamo is justified based on the imaginative geography of formerly colonized areas. Not only are representations of those imaginative geographies derived from colonialism, but the concrete geographical and historical background of the camps also make this place a space of exception (Reid-Henry 2007).

In comparison to cases in Western countries, there are relatively few studies focusing on those in East Asia. From a historical perspective, in modern East Asia, the control policy of migrants, including irregular ones, had been shaped by Japanese imperialism, its vestiges, and the influence of the US during the Cold War. In particular, the migration flows from the Korean peninsula to Japan before and just after the end of World War II (WWII) share several characteristics that concern the relationship between islands and incarceration. While the islands between Japan and the Republic of Korea (ROK) became the main stage of post-war border (re-)demarcation as the frontier of these countries (including the disputed territories of Takeshima or Dokdo) after the war, the Tsushima/Korea Strait (Fig. 1) was an active sea area for the passage of people and goods when Korea was annexed by Japan. Going further back in history, from the late 19th to the early 20th century, the islands on the Japanese side were recognized as its frontier security bases against threats from Russia. The

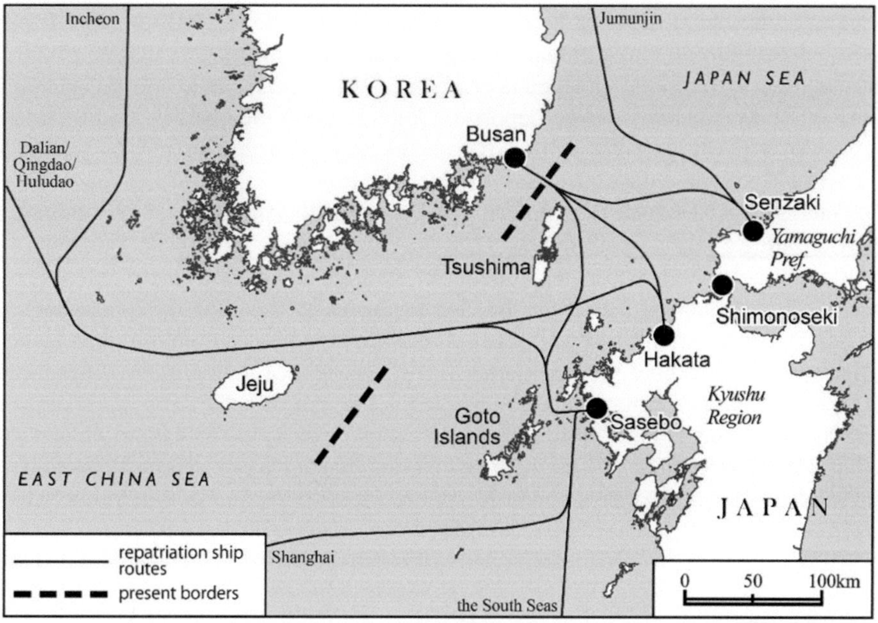

Fig. 1 Major ports designated for repatriation and repatriation ship routes across Tsushima/Korea Strait

fracturing nature of marginality and centrality of these islands is noteworthy when considering the irregular migration of Koreans during the US occupation.

In addition, the Japanese government attempted to treat such irregularized migrants in a colonial manner reminiscent of past history, although its sovereignty had been restricted by the U.S. occupation force. We should not regard the sovereignty of superpowers as an inherent condition when examining the perpetuation of the "colonial past" (Martin and Mitchelson 2009). If sovereignty is reconstructed through incarceration, how did sovereignty in colonialist immigration control persist? This chapter focuses on the specific process of (re)construction of Japanese sovereignty through the analysis of "spaces of exception" in Sasebo, with the geopolitical changes of the islands in mind.

Through the analysis of historical materials, I pay special attention to the dialectics and 'micropolitics' of inclusion and exclusion of minority populations, one of the most salient characteristics of colonialism. Based on the works of Agamben, Takahashi (2011) argues how irregular migrants or rebels are easily stripped of basic human rights via state sovereignty and are separated and marginalized from authorized nationals, while they are simultaneously viewed as biopolitical entities and subjected to the sphere and regime of sovereignty. In other words, while detainees are the ones exposed to what Agamben refers to as "bare life," migrants themselves, including those subjected to sovereignty protocols, are the ones enabling and performing extralegal practices. We can observe these two sides of sovereignty in the context of the Japanese Alien Registration Law. From 1955 to 1993, this legislation required former colonial nationals to register their fingerprints; as discussed later, the Japanese government intended not only to identify and separate "illegal" persons from other Koreans but also to deport the latter, even if they had resided in Japan before WWII. Unless Koreans in Japan could prove their legitimacy to stay, their existence would be reduced to "bare life" in detention camps.

The rest of this chapter is organized as follows. In the next section, the situation of Korean "illegal" entrants during the U.S. occupation period will be summarized in relation to legislation concerning "liberated" Koreans and the locational characteristics of Sasebo. Section 3 explains the circumstances surrounding the establishment of a detention camp in Sasebo. In Sect. 4, historical materials highlight the characteristics of the Sasebo detention center as a space of exception and the differences in the treatment of entrants regarded as illegal and that of authorized repatriates to Korea. The features of the Sasebo detention camp are analyzed in Sect. 5, which also examines the survival of sovereignty in a colonial manner, paying special attention to the identification and suppression of Korean residents in Japan. Section 6 concludes.

2 The Control of Koreans' Mobility Before and After WWII

The strait area between Japan and the Korean Peninsula underwent a transformation in step with the advance of Japanese colonialism, as described above. During the early modern period, the Russian threat to Japan was significant enough to warrant the establishment of military facilities, such as watchtowers, on islands scattered across the strait. In other words, as a Japanese periphery, the military strategy greatly contributed to the hardening of the borders around the strait. In this process, Sasebo was given an important role in the management of these islands. Although Sasebo was once an impoverished village before the modern era the Japanese Imperial government decided to establish a naval district responsible for the Korean Peninsula and the Ryukyu Islands in 1889 because it had the natural conditions for a good port. As the threat from Russia intensified, watchtowers were built on the southern islands of the peninsula and on Jeju Island (later annexed to Japanese territory), Tsushima and the Goto Islands (Fig. 1). These watchtowers were linked to Sasebo by undersea telegraph lines. In a sense, Sasebo served as a hinterland for boundary demarcation in the region. After the annexation of Korea, many passenger and cargo ships passed through the Tsushima/Korea Strait. With the disappearance of military threats, the sub-branches of the Sasebo Naval District that managed facilities on those islands were closed. Consequently, this sea area became less isolated.

Just after the end of WWII in August 1945, approximately two million Koreans in Japan, including those who were forced to migrate as conscripted labor, were eager to return to Korea. This eagerness led some of them to rush to the ports to board ships back to their homeland. By April 1946, the number of repatriates had reached

1.4 million, who had reentered Korea either on their own or through the official transportation plan implemented by the occupation forces (Wagner 1954). Although almost 90% of the remaining Koreans wished to repatriate (Division of Social Affairs, Ministry of Health Affairs n.d.), the number of repatriates barely increased after 1946, as the unstable political and economic situation in the Korean Peninsula stimulated migration or reentry into Japan. At the same time, the increased transit in the opposite direction was prompted by individuals who had already returned to Korea. In several cases, family members were separated between the two countries because a part of the family had been sent solely to confirm living conditions in their homeland. In short, most of the irregular migrants were individuals seeking to reunite with their families. During the occupation period, over 47,000 irregular migrants were reportedly captured at sea or on land (Morita 1996: 112); however, the actual number is probably much higher.

In recent studies, the reconstruction of social boundaries through legislation during the occupation period has been actively discussed in relation to the oppression of irregular migrants from Korea and their deplorable treatment in detention camps. As early as the spring of 1946, some local military government teams firmly recognized that the influx of irregular migrants, most of whom arrived by sea at Tsushima Island or coastal areas of the northern Kyushu Region and the western Chugoku Region (Fig. 1), was detrimental to the implementation of occupation policies. Consequently, they deported them based on the respective government's decision. For example, an Australian officer in the Yamaguchi Military Government Team[1] requested the General Headquarters of the Supreme Commander for the Allied Powers (GHQ/SCAP) to prepare an ordinance to identify the "illegality" of the entrants because local officers had few means to distinguish them (Morris-Suzuki 2010). The absence of such resources was also challenging for individuals entering the land. In Osaka, which had the largest Korean population in Japan, the prefectural government issued a registration regulation (exclusively targeting Koreans) in November 1946 to determine the status of "illegal" entrants.[2] Moreover, ordinary Japanese people, as well as local and military authorities, engaged in the suppression of irregular migrants. In several prefectures, local military authorities built surveillance systems to monitor the arrival of entrants, with the assistance of local police and residents (Ri 2023); their actions created negative imaginative geographies that impacted irregular entrants. The islands and these coastal areas became peripheral zones of Japan. It was under such circumstances that the Repatriation Relief Bureau was established in Sasebo. It was converted from a naval facility and became the focal point for movements of Japanese and former colonial subjects.

In considering the oppression faced by irregular Korean migrants and the demarcation of borders, it is important to take into account the historical changes in the legal status of Koreans. Japanese colonial rule, especially in Korea, exhibited many

[1] British Commonwealth Forces handled occupation policies in the Chugoku and Shikoku regions.

[2] In this regulation, the police played an important role in conducting registration. However, GHQ/SCAP had a negative stance towards police involvement in non-criminal areas, which resulted in the withdrawal of the regulation during the preparation for the Alien Registration Ordinance.

characteristics similar to those in other colonized territories. The Japanese government applied a different legal system to its colonies than that used in mainland Japan (*Naichi*). Particularly, the legislation of mobility control and citizenship is a salient characteristic of the Korean administration. Although Koreans had Japanese nationality, their status was strictly distinguished from that of native Japanese based on family registration; Koreans were not allowed to transfer their register to Japanese individuals, and vice versa (except for intermarriages and the adoption of children).Their free movement from the Korean Peninsula to the Japanese mainland was restricted; they had to obtain a permit issued by the colonial authority (Busan port was effectively an immigration inspection center). Furthermore, even after entering mainland Japan, Koreans were required to submit a notification when they relocated to their residences. Moreover, Koreans had no right to renounce their Japanese nationality because the government intended to exercise its sovereignty over independence movements carried out abroad.

After the war, entry to and exit from Japan without the express permission of the occupation forces was prohibited; therefore, irregular movements in the early stages of the occupation were treated as "illegal" for violating the established order. Simultaneously, the legal status of former colonial citizens was quite ambiguous: while they were initially treated as "liberated people"—and not as either Japanese or as foreigners, as determined by the Allied Powers—the occupation authority decided to render these individuals subject to Japanese criminal jurisdiction in March 1946 due to concerns about their alleged engagement in the black economy and communist activism. Moreover, without clearly determining their nationality, these individuals were forced to register as "foreigners," in line with the Alien Registration Ordinance of May 1947. Since foreigners of the Allied Powers were excluded from the application of this ordinance, it was, in effect, those from the former colonies, especially Koreans, who were specifically targeted. This Ordinance contained an article concerning the deportation of these foreigners, even though they were subject to Japanese criminal or administrative laws. This legislation was a form of collaboration between GHQ/SCAP and the Japanese government, enabling the administrative authority to arbitrarily determine what and who is considered "illegal." Morris-Suzuki (2010) identified a continuation of colonialism in this context, which was evident alongside the demarcation of national borders.

Although few studies have analyzed irregular Korean migrants before the 2000s, an increasing number of historians and sociologists have actively investigated this and other related topics in recent years. However, the practices regarding legality and illegality, and their dialectics, have been generally understudied, particularly in relation to legislation and border controls during the U.S. occupation period. The continuity of colonial power, especially its dialectics of exclusion and inclusion, was vividly evident in the detention camp in Sasebo, despite being operational for just four years.

To investigate this issue, this chapter mainly relies on two sets of historical documents: *The History of the Sasebo Repatriation Relief Bureau* (SRRB 1949, 1951), published by the institution that managed and operated facilities such as the detention camp; and *Keiko*, a journal published to promote friendship among police officers in Nagasaki Prefectural Police, which was responsible for guarding the camp under the instruction of the Sasebo Military Government Team.

3 "Illegal" Entrants in the Sasebo Detention Camp

3.1 The Opening of the Sasebo Detention Camp

This section summarizes the establishment and initial operations of the Sasebo Repatriation Relief Bureau (hereafter SRRB), which was the only institution dealing with the deportation of irregular migrants. Sasebo was originally designated as one of the ports to manage repatriation operations from and to Japan. This involved implementing quarantine measures and coordinating with local occupation forces. Sasebo was considered suitable for these operations because of its numerous naval facilities located on Hario Island (Fig. 2) and barracks that could be converted to accommodate repatriates. Approximately 40,000 "illegal" entrants were deported through this port between 1946 and 1950.

Return migration from the Korean Peninsula became prominent at the beginning of 1946, and occupation authorities gradually began to regard it as a problem. According to the order issued by the occupation force in March 1946, returnees were prohibited from traveling to Japan without permission (Kim 1997: 263). Therefore, the authorities created a category of "illegal" migrants who were to be treated differently from "ordinary" repatriates. Although no specific procedures for deportation were stipulated at this time, some migrants had already been deported directly from Hakata and Senzaki, two ports designated for repatriation, in February of the same year (Morris-Suzuki 2010). The GHQ/SCAP, for the first time, clarified how to deport these individuals on June 21, 1946, based on SCAP Instruction No. 1015 ("Memorandum on Suppression of Illegal Entry"), mainly driven by the need to prevent the influx of cholera. This memorandum stipulated that illegal migrants captured by local authorities had to be deported from Hakata, Senzaki, Sasebo, and Maizuru in Kyoto Prefecture showed in Fig. 1 (however, since vessel traffic with Korea had ceased in Sasebo at this point, deportation was not actually taking place).

In response to these instructions, the military government in Sasebo ordered the prefectural police to prepare to deal with "illegal" entrants. For example, the police set up coast guard units at 88 locations to monitor the entry of suspicious migrants, while the Hario Guard Unit (consisting of some 100 police officers) was responsible for deportees' security. Barracks No. 6 and No. 12 were also converted to temporarily host detainees (Security Division, Nagasaki Prefectural Police 1950; Fig. 3).

However, from July to August, the deportation system changed drastically: for instance, irregular entrants were to be sent only through the SRRB. The specific reasons for this change are unclear, except for the limited capacity to accommodate cholera patients in Senzaki (Suzuki 2006) due to a dearth of related documents. What is clear is that the consolidation of deportees in Sasebo was achieved within a short

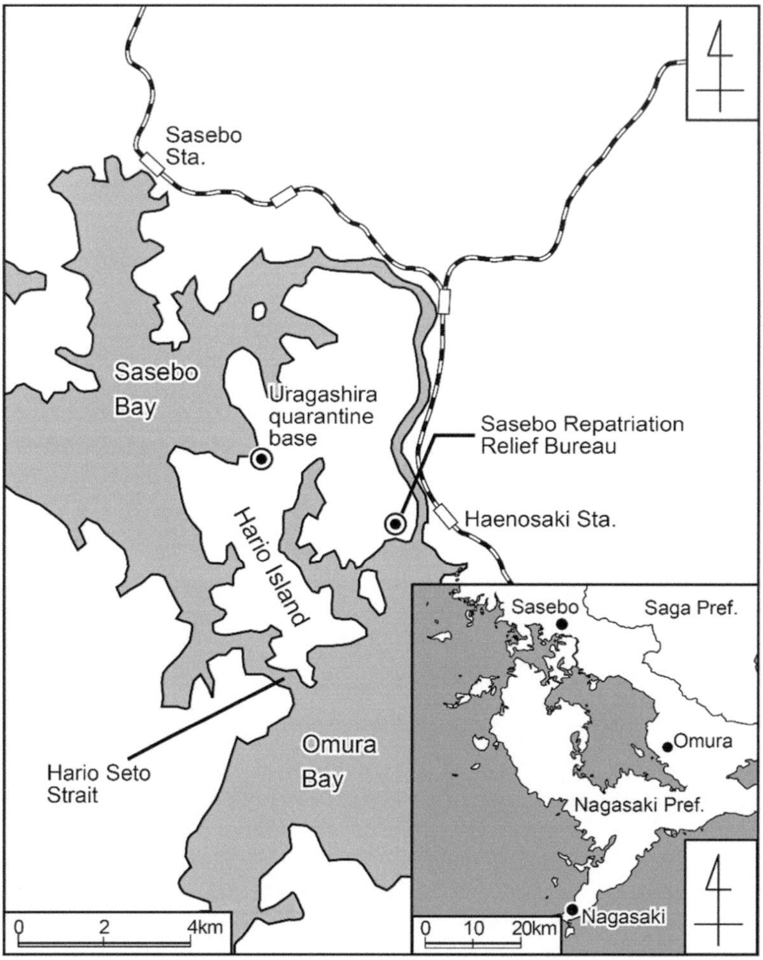

Fig. 2 Location of the Sasebo Repatriation Relief Bureau and its related facilities in 1947

period of time. The decision to make Sasebo the main entry point came as a surprise to police officers and civilian personnel stationed there. For instance, a local security official expressed his bewilderment as follows:

> It is surprising that, from this time onward, Sasebo is assigned as the only place handling Korean illegal entrants, previously sent to Senzaki and Maizuru (Yoshimura 1963c: 51).

Shortly after the aforementioned change, the number of entrants greatly exceeded initial expectations, and it became difficult to accommodate deportees, including those suspected of being illegal entrants, in the existing facilities. During early August, approximately 7,000 Koreans were held in transport vessels or small towing ships in Sasebo Bay, and more than 3,000 persons were incarcerated on land. Consequently, the Relief Bureau was forced to convert Barracks No. 9 and No. 11 into

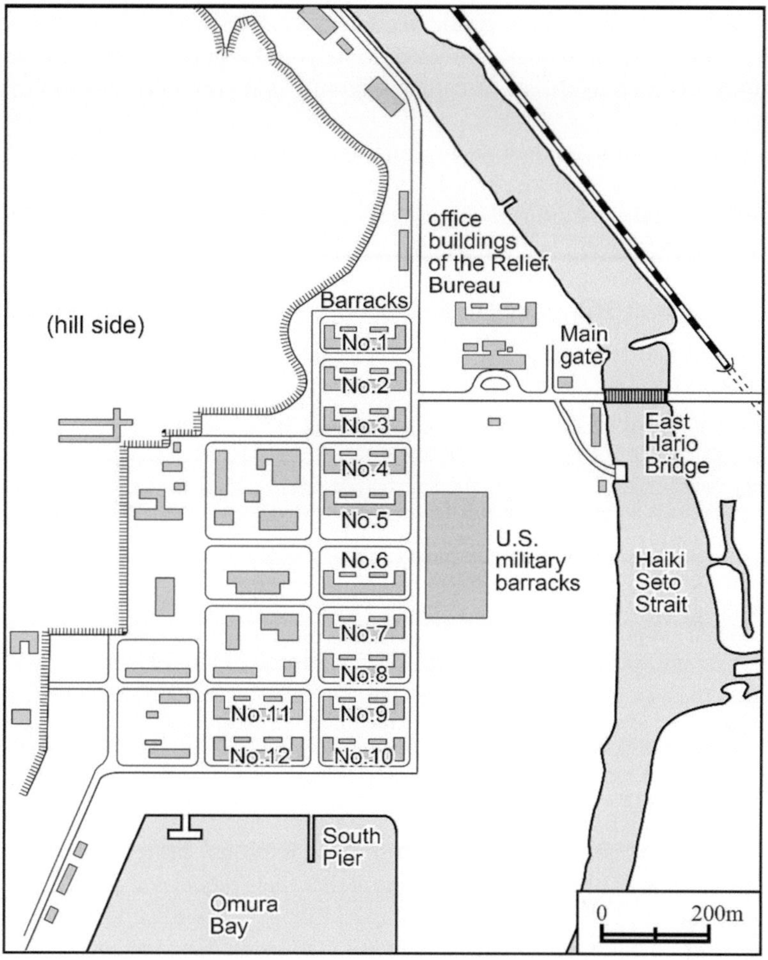

Fig. 3 Map of the Sasebo Repatriation Relief Bureau. 1949 *Source* SRRB: 55, redrawn by the author

detention camps, while the prefectural police increased the number of officers in the Hario Guard Unit to 230 (Security Division, Nagasaki Prefectural Police 1950). A deportation system was gradually established in Sasebo, and more than 15,000 people were deported back to Korea over the next four months until December 1946.

One of the characteristics of deportation during this time was that the duration of detainees' stay in Sasebo was considerably longer than in later periods. For example, a transport ship that arrived in Sasebo on its way back to Senzaki, as mentioned earlier, was finally allowed to leave the port in early September, and in one case, due to a railway strike in Busan, a deportation ship was detained for nearly two months (Yoshimura 1963a). The main reason for this prolonged detention was the strengthening of the quarantine system to prevent the spread of cholera. In other words, owing to the incubation period of cholera, it was necessary to isolate patients or persons who were potentially infected. Moreover, because there were no cholera quarantine facilities in Busan, irregular migrants had no choice but to remain in Sasebo.

However, even for quarantine reasons, the conditions under which detainees were placed, especially those aboard ships, were appalling. Of course, with respect to cholera quarantine, Japanese repatriates from abroad were treated the same as Koreans or other repatriates; nevertheless, deportees were not allowed to move to hospitals on land or hospital ships. The following excerpt from *Keiko* journal suggests dissatisfaction with the local authorities and offers interesting insights for subsequent analyses:

> September 4 [1946]
>
> A representative of the detainees isolated in one ship: What crimes are we charged with? Explain the degree of punishment, disposition, etc., in detail.
>
> Police officers in the Guard Unit: You are charged with violating the orders of the Allied Powers, and we do not know the extent of the punishment. However, the local military government ordered anyone who rioted or fled to be shot.
>
> [...]
>
> Representative: When will our internment be finished, and if there is a cholera outbreak, why do you not place healthy people in a separate room to prevent them from becoming infected, or just send them to Korea early? Is this an order of the Allied Powers?
>
> Answer: You have no choice but to endure the circumstances until internment is finished. Those who have lived in a house where cholera appears are treated as carriers of this disease. Even police officers here, when replaced, have to stay in quarantine for a week, and only when they are examined and found to be in good health are they permitted to land. We cannot make exceptions (Yoshimura 1963a: 14).

Even with the risk of cholera infection, the Koreans' protests were justified. However, in the end, there was no indication that their requests were heeded. After the consolidation of illegal entrants in Sasebo, their initial treatment and deportation were carried out within an atmosphere of confusion. Simultaneously, the above document highlights the source of power through which the incarceration of irregular migrants occurred. This power prevented Koreans from escaping and enabled the authorities to shoot those who attempted to flee.

3.2 The Sasebo Detention Camp as a Space of Exception

The transfer of entrants (recognized as illegal in some cases) from various parts of the country was conducted by railroad or sea (with detainee ships being towed by transport vessels or small boats), as mentioned earlier. The local police at the place of departure were in charge of security throughout the transfer process until the detainees reached Sasebo, where they were handed over to the Uragashira Quarantine Base (in the case of sea vessels) or to the security officers of the Hario Guard Unit at Haenosaki Station (in the case of land vehicles) (Fig. 2). After the cholera outbreak was under control, detainees were transferred to Barracks No. 12 by truck or by other means. Finally, they were transferred by sea to a ship in Sasebo Bay through the narrow strait located on the west side of Hario Island (Hario Seto), accompanied by security police officers, to Busan and handed over to local officials upon arrival. Thus, illegal entrants were under constant police surveillance from the time of their arrest until they were landed in Busan.

What is noteworthy about detainees' treatment is that the physical environment of the barracks reinforced this surveillance system. Even in the early stages of official repatriation, the dormitories of Korean and Chinese detainees were set up in a manner that separated them from Japanese repatriates, as noted in the following excerpt from *Keiko* journal:

> The dormitories for Koreans and Chinese were surrounded by barbed wire and interaction with other dormitories was prohibited.[3]

However, as shown below, after the deportation of irregular migrants began, the barracks for detainees were modified to be more strongly isolated from the exterior.

When Sasebo was designated as the deportation port by SCAPIN 1015 in June 1946, Barracks No. 12 were modified. They were:

> [T]emporarily reformed and divided into eight rooms to accommodate up to 1,000 people, while latticework made of lumber was added to the walls and entrances of the rooms and the camp, *like a detention cell in a police station* (Security Division, Nagasaki Prefectural Police 1950: 30) [emphasis added].

The detainees were not only treated differently from official repatriates but were also kept in isolation and under surveillance as criminals. The construction of physical barriers outside the barracks was confirmed in a report by the chief officer of the Guard Unit.

> September 18 [1946]
>
> Today, there has been a change in the officers of the U.S. Army in charge of directing us [police officers]. The newly appointed Captain F. M. Combs gave us the following instructions:
>
> 1. After sending detainees from Barracks No. 1, 9, 11, and 12, only No. 12 will be used for illegal entrants, and the rest for repatriation to Okinawa. Barracks No. 12 will have double-barbed wire netting and better lighting to improve security. I ordered this to the Relief Bureau.
>
> 2. […]
>
> 3. Detainees who flee or riot may be killed at your [the Guard Unit officer's] will. The authority of Colonel Bean, the regimental commander, was assigned to the Chief.
>
> 4. Police officers at the site are under our command. All of you are responsible for vigilance against illegally migrating Koreans and ensuring that there are no omissions (Yoshimura 1963b: 42).

---

[3] Division of Mainland China, Department of Overseas Japanese, Bureau of Administration, Ministry of Foreign Affairs, "Landing conditions of Japanese repatriates (in Japanese)," March 1, 1946, owned by the Diplomatic Archives of the Ministry of Foreign Affairs in Japan.

Together with the installation of additional barbed wire, the "Directive dated September 30 (from the U.S. Army Commander in Hario to the deputy director of the Bureau) ordered more steel wire to be set up on the barbed wire mesh around Barracks No. 12,"[4] indicating that serious attention was paid to the separation of the detention camp from the rest of the facilities. Furthermore, the landscape in and around the camp was transformed by the placement of guards and a police observation box on the mountainside, west of Barracks No. 12.

The camp was modified thoroughly to prevent detainees from escaping. However, even after several layers of barbed wire were installed, there seemed to be no end to the constant escapes. The camp originally served as a barracks and was not suitable for detention. Thus, efforts were being continuously made to thwart escapes, as stated in the following report:

> Recently, the inadequacies of the barracks have been improved, and the board fences have been renewed, making it impossible for them to escape (Seki 1950).

This microscale reinforcement of spatial boundaries was accompanied by the extralegal treatment of illegal entrants, as shown in the above excerpts. In short, it was permissible to immediately open fire on detainees without any special procedures, as is evident in the "should be shot" or "may be killed" wording. Although it is not clear whether anyone was actually shot at and/or killed, exceptional measures were certainly feasible, such as firing warning shots to suppress disturbances over gambling in the camp (Yoshimura 1963b). Thus, various practices to block off the camp and detainees' unreasonable treatment are interrelated, and in this sense, Barracks No. 12 was indeed a space of exception.

These spaces were created through local military governments' proactive roles. As mentioned above, Sasebo's police security system was established by the military government. In fact, the prefectural or local police did not engage in the operation of the barracks: the chief officer of the Guard Unit confessed that.

> All ships [transporting the detainees] were under the command of the occupation force, and [they] simply [moved] based on its directives. [They] did not know when or where the ships would arrive (Yoshimura 1963c: 52).

In other words, the space of exception in Sasebo embodies the power of the occupation force.

Although considerable evidence points to the exceptional treatment of detainees, the nature of power in the space of exception in Sasebo appears clearly in documents written by Japanese staff or officials in the Repatriation Relief Bureau or the Unit. For example, a staff member of the Bureau who reported the daily admittance and deportation of illegal entrants by officers in the military government expressed his fear of their exceptional power as follows:

---

[4] From Assistant Consul Nishikawa, Liaison Group in the Sasebo Repatriation Relief Bureau and Sasebo Branch of the Central Liaison Office, to Director General Yamanaka, Bureau of Administration, Ministry of Foreign Affairs, "Subject: U.S. Military Command on the improvement of facilities and repatriation affairs in the Sasebo Repatriation Relief Bureau (in Japanese)," October 11, 1946, owned by the Diplomatic Archives of the Ministry of Foreign Affairs in Japan.

> [With respect to the reports on their number] It is a huge misunderstanding if you think that the officers fail to notice or care about it, even if you get one or two persons wrong. The U.S. military is punctual in everything and never lets you off the hook for just a slight error. If they found an error in the report, they said that if we made another error, the person in charge of the liaison operation would be placed in Barracks No. 12. I could not help turning blue (Kagino 1951: 228).

Similarly, police officers guarding the barracks were under strict orders to prevent detainees from fleeing and were forced to work hard:

> As night fell, the wind and rain increased. There were no boxes, and the lights were going out, so every officer was tense for fear that the Koreans in Barracks No. 12 might escape. They were exposed to the rain for three hours. I [the chief of the Guard Unit] cannot thank my subordinates enough (Yoshimura 1963a: 12).

The police felt considerable pressure from the military government, as evidenced by reports such as this:

> The trouble at Barracks No. 12 last night was so repugnant to the occupation force that the commander himself came out and issued a stern warning through Sergeant Ninomiya (Yoshimura 1963c: 52).

Under these circumstances, some of the newly assigned police officers ran away.[5] Even more interesting is that the military government recognized and acknowledged that Barracks No. 12 was a space that differed from the others.

> It had been more than two years since the issuing of the restrictions on entry [of newspaper reporters] to the Bureau at the beginning of Showa 23rd year [1948]. As it was pointless, they rarely approached here, and even the Bureau staff had forgotten about these restrictions. Currently, the number of repatriates has considerably decreased, and we think the extreme restrictions seem useless. Therefore, we asked the local military government about handling reporters entering the Bureau. The government answered the following.
>
> 1. Never wander around or inside Barracks No. 12.
> 2. Do not photograph the U.S. Military Camp and Barracks No. 12.
> 3. Strive to select articles that are not detrimental to the performance of repatriation operations.
>
> Just these three conditions were proposed, and their entry was fully permitted (SRRB 1951: 180).

The juxtaposition of the U.S. military camp and the detention camp as exceptions for permission is particularly noteworthy here. These two spaces were placed under the extralegal authority of the occupation forces.

Unfortunately, there is insufficient detailed information on the military government's perception of the creation of these spaces. However, one possible explanation is that the SRRB regarded irregular migrants as major impediments to implementing occupational policies. Numerous references to illegal entrants exist in the documents created by local staff and police officials. Notably, there were several protests from

[5] The harshness of their work is seen from the memoirs recorded on *Keiko*: "I missed the lottery for today's transfers. I am wet and crying in Barracks No. 12. I hate the fact that my friends left" (Ishibashi 1950). This was expressed as a tanka poem.

detainees expressing their dissatisfaction with their treatment and the camp's conditions regarding rations, gambling, and other issues. In response, as in the cholera case discussed above, local authorities strengthened their control over the detainees without adopting effective remedial measures. The detainees' exceptional treatment was strongly motivated by the local military government's intention to maintain order through coercion.

4 The Division Between Illegal Korean Migrants and Official Repatriates

The SRRB handled both deportation and official repatriation abroad from the beginning of 1947—when the cholera outbreak had almost subsided—to May 1950—when the Bureau closed. This section highlights that the distinction of "illegal" Koreans should be understood in relation to the legitimacy of Koreans in Japan, especially vis-à-vis the proclamation of the Foreign Registration Ordinance and its related discriminatory practices.

Several differences were observed between both types of returnees during this period. In addition to illegal Koreans being isolated in Barracks No. 12, there were other differences. For instance:

> From December of the same year [1947], it was established that Japanese and foreign repatriates would receive 600 grams of staple rice per day, while deportees would receive 500 grams. Consequently, the quantity and quality of food has greatly improved (SRRB 1951: 137).[6]
>
> As there is a difference in food rations between official repatriates and deportees, we sometimes receive complaints from the latter. On the ship, we have representatives from each group, and we cook [the rice] in separate pots (SRRB 1951: 221).

Different practices related to freedom of movement should also be emphasized. While illegal entrants were always under police surveillance to deter them from escaping, some official repatriates attempted to leave the Bureau. For this reason, the Bureau set up a watch station, while civil personnel from the Security Affairs Division of the Bureau were in charge of monitoring (SRRB 1951: 178). Albeit under quite limited conditions, official repatriates were permitted to go out:

> In principle, the repatriates were prohibited from going out, but when there was a truly unavoidable reason, such as to check luggage or go to the hospital, the chief of the Bureau could allow applicants to go out once a day in a group (led by a responsible person) under the approval of the occupation force (SRRB 1951: 178).

Thus, the movement of official repatriates was not completely prohibited. Additionally, they were not under police surveillance. Furthermore, although both groups had boarded the same vessel to Busan:

---

[6] However, Korean detainees were thought to have initially received more rations than the Japanese, as they were treated as any other foreigners, as Yoshimura (1963a) indicates.

> [N]ot only were regular repatriates and deportees not allowed to mix on a transport boat (between the Akago Pier and a vessel on Sasebo Bay); [but] security concerns made us carry only limited numbers of deportees at any one time (SRRB 1951: 162).

Transportation from the Bureau was conducted separately. Figure 4 shows a boat at the Akago Pier, where police officers monitored the boarding of deportees. They were kept under constant police surveillance and subjected to extralegal powers until they were handed over to the local authorities in Busan.

Notably, in these processes of deportation, the local authorities made efforts to verify the "legitimacy" of ordinary repatriates and the "illegitimacy" of the rest. The legitimacy of official repatriates was verified via certifications issued by national or local governments. When the repatriates arrived at Sasebo from various places in Japan, the local authorities first checked their certificates of repatriation issued by a local government of their departure municipalities in order to prevent repatriates mixing with "illegal" persons who "bought certificates for repatriation from those who had abandoned their intention to return home and attempted to enter the Bureau" (Otao 1949: appendix p. 20). These means were supposedly resorted to by "illegal" entrants who had succeeded in landing in Japan but could not return to their homeland regularly, even if they had special reasons to do so (Hyakutake 1951).

However, especially after the enactment of the Alien Registration Ordinance on May 2, 1947, the verification procedure began to extend to registration cards, and it gradually became impossible for Koreans to return to their homeland other than

Fig. 4 Photograph titled with "Boarding of deportees at Akago Pier". 1951 *Source* SRRB: appendix p.8

through official repatriation. In Sasebo, based on a directive from the GHQ/SCAP on December 23, 1947, it was prescribed that:

> When Korean deportees arrive in southern Korea, appropriate measures should be taken to ensure that they do not have official documents and certifications, such as alien registration cards or ration coupons, whether issued by official institutions or forged, to avoid giving them a chance to reenter Japan (SRRB 1951: 75).

In particular, the staff of the Bureau conducted strict checks on foreign registration cards, as shown below.

> In general, their alien registration cards had to be returned when they left Japan to prevent unauthorized use of the cards that enabled illegal entry or smuggling, and then the collected cards were sent back to the municipality offices where they were issued. However, in agreement with the Ministry of Justice, we temporarily held the cards for verification, and the certificates were carefully inspected. If we found any discrepancies in content, unclear tally impressions, or evidence of photo replacement, we inquired about the municipalities that originally issued the cards and approved the repatriation subsequent to valid replies.
>
> In Showa 24th year [1949], the Alien Registration Ordinance was revised to make quasi-visa operations relatively easy (SRRB 1951: 105).

These verification processes served a function in the identification of legitimate or illegitimate persons, as shown by the following statement:

> Of the 115 Koreans scheduled to board Shinsui Maru, which departed on February 6, 1950, we inquired about the validity of their foreign registration cards with each municipality, and 13 were found to be illegitimate repatriates. All 13 feared arrest and fled" (SRRB 1951: 105)

The Alien Registration Ordinance and its subsequent revisions increased the certainty of the verification. In the confirmation of the initial certificates of repatriation, there was no choice but to make judgments based on written data, such as age, gender, and name; this led the personnel of the Bureau to be often confused due to mistakes by the municipalities that issued certifications (Otao 1949). However, after the proclamation of the Ordinance, "Since the repatriates were required to bring their foreign registration cards with a photograph attached, such fraudulent entries [to the Bureau] were almost eliminated" (Otao 1949: appendix p. 21). As discussed in the next section, the 1949 revision, which included the implementation of a nationwide serial number, streamlined the system for issuing registration certificates.

The Foreign Registration Ordinance's identification practices indicate that even ordinary repatriates, regardless of their actual migration experience, were to be treated as illegal entrants at risk of being subjected to extralegal treatment unless they could demonstrate their legality. Moreover, the significant role played by the Ordinance legislated by the national government shows that microscale practices to create a border between legitimacy and illegitimacy were linked to the (re)construction of spatial and social borders at the national scale.

5 Fixing the Boundary Between Legitimacy/Illegitimacy and a Lingering Colonialism

Although prior studies have long problematized the Alien Registration Ordinance within the context of the deprivation of Koreans' civil rights and the exclusion of former colonial nationals, it must be understood as the source of Korean immigrants' illegitimacy status. However, when the Ordinance was first enacted, the identification of illegal suspects based on registration cards proved inadequate. This was because, in some cases, bulk registration through the League of Koreans[7] was still permitted (as the registration rate remained quite low without assistance), resulting in double registration and the trade of registration cards.

Some police officers considered it impossible to immediately identify the holder of a registration card as a legitimate resident (Fukumoto 2008). And so, various efforts were made to make identification based on registration cards more effective. For example, the Kyoto Prefectural Government checked registration cards using a ration passbook in 1948 (Kyoto Branch of Central Liaison Office 1948). The December 1949 revision of the Ordinance prescribed a valid period of cards (three years), the mandatory carrying of registration cards at all times, and severe punishment (as a criminal offense) for individuals without cards, in addition to the implementation of a nationwide serial number (Kim 1997: 666–667). From the beginning of promulgation of the Ordinance, the Japanese government emphasized the authenticity of registration in order to identify illegal entrants based on the Ordinance (Chung 2012: 127). However, the supplementary practices proposed after the enactment of the Ordinance aimed not only to identify illegitimate entrants more effectively but also to solidify the boundary between card holders and non-card holders among Koreans in Japan.

Here, we should analyze the arbitrary creation of the aforementioned boundary and the exceptional power that made it possible. These functions of the Ordinance implicitly appeared in its definition of "persons without legal residence." The term "without legal residence" referred mainly to individuals who did not have a rice ration passbook. During the first half of the U.S. occupation, it was difficult even for internal migrants in urban areas to obtain a regular ration passbook because they were restricted by the authorities due to food shortages, except for individuals who were repatriated from abroad and had returned from evacuation. Therefore, even Koreans who had lived in Japan before the war were at risk of being identified as individuals "without legal residence" and subject to deportation, regardless of whether or not they were irregular entrants.

For Koreans, the only way to avoid the risk of deportation was to procure a registration card. However, the Alien Registration Ordinance contained a provision (in Articles 13 and 16) that allowed for the deportation of those who sold or transferred cards and those who failed to carry their card at all times. Simultaneously, shortly

[7] This association organized the largest number of Koreans in Japan during the U.S. occupation and had a strong leftist leaning. It was ordered to dissolve over breaches of the Organizations Control Order in 1951.

after the enactment of the Ordinance, the Japanese government stipulated that the municipalities were to issue a "provisional certificate" different from those of regular applicants if a person "without legal residence" applied for registration (Chung 2012). The validity the provisional card lasted three months, and the carrier was kept under the supervision of local authorities until he/she could prove his/her legitimacy. In any case, these Koreans continued to be at risk of deportation, regardless of whether they complied with the legal system or not.

Ultimately, the enforcement of deportation based on the Alien Registration Ordinance deviated from the normal administrative system of immigration control in that even potentially legal residents could be subject to deportation. This extralegal and arbitrary creation of the legal/illegal boundary was equivalent to the (non)possession of registration cards.

In light of the Ordinance's features, the various measures taken against Koreans in Sasebo imply the existence and connection between two types of boundaries. First, concerning physical boundaries, various modifications were made to Barracks No. 12, including the installation of barbed wire and watch boxes to prevent detainees' escape, which clarified the separation of the detention camp from the exterior and eliminated the possibility of crossing this micro border. Second, there were practices to create and sustain the social borders of legal/illegal persons through, for example, the recovery of rice passbooks or the verification of foreign registration cards in a manner unique to the Bureau of Sasebo, as discussed in the previous section.[8] Local authorities sought to eliminate any room for illegal elements to disturb the boundaries determined a priori. Thus, the Sasebo detention camp was more than just a space where irregular migrants were suppressed. Exceptional powers imposed physical restrictions on illegal entrants: containment in the camp, deterrence of escape, constant surveillance, and spatial isolation. These acts were combined with the legitimacy/illegitimacy division based on verifying authorized documents. Consequently, the boundary between illegal entrants and regular repatriates materialized as meaningful and unbreachable.

Illegal entrants were not the only group subjected to exceptional power. To avoid incarceration, official repatriates had to identify themselves as legitimate by possessing certain documents. To prevent the breach of physical boundaries, repatriates and detainees were both subjected to exceptional powers.

The essence of Koreans' treatment under the U.S. occupation was materialized in the Sasebo detention camp. That is, if entrants were determined to be illegal, based on the arbitrary boundary between legitimacy and illegitimacy, the authorities were allowed to treat them using extralegal means and force all Koreans to submit to this categorization. While this exercise of powers was made possible by the acceptance or guarantee of GHQ, we can identify the persistence of colonial rule, or the dialectic

[8] For security officers in Sasebo, illegal entrants and persons "without legal residence" were synonymous. One of them said, "With the recent revision of the Alien Registration Ordinance, the number of persons without legal residence, or so-called *illegal entrants*, is expected to increase compared to last year. This situation makes the Hario Guard Unit even more important" (Nishioka 1950) [emphasis added].

of exclusion and inclusion discussed earlier, in the microscale practices occurring in the Sasebo detention camp.

However, the arbitrary creation and enforcement of boundaries did not succeed unilaterally. There was significant opposition to the enactment of the Alien Registration Ordinance from major Korean organizations in Japan, especially since the mandatory carrying of registration cards at all times was reminiscent of the prewar police oppression of Koreans. However, the central headquarters of major Korean organizations eventually accepted the enforcement of the Ordinance because it appeared that the occupation authorities would, to a certain extent, recognize "legitimate" ordinary Koreans as foreigners and accommodate the requests from these organizations (Kim 1997: 363–375). Although there were some complaints about the deportation of persons "without legal residence" or desires to exclude demobilized Koreans from these persons at least (Chung 2012), few objections were raised against the Ordinance article that enabled the deportation of Koreans (Kim 1997). Additionally, some Korean media in Japan criticized the illegal entrants themselves. For example, the editorial in the *Chosen Nippo* [Daily News of Korea] on September 28, 1947, denounced the irregular entrants for exacerbating the chaos in Korea and providing handy excuses to the authorities to implement the ordinance (Kobayashi 2011).

Kobayashi (2011: 124) argued that "Korean media in Japan tried to fend off pressure for deportation [against the trend to expel all Koreans] by emphasizing that they were also victims of illegal acts committed by some Koreans," and interpreted this as resistance to the authorities. However, this interpretation seems inaccurate, since this criticism is spontaneously directed at irregular migrants without considering that even ordinary Koreans themselves would be subjected to extralegal requirements, such as the obligation to carry and present alien registration cards at all times and to be deported. In other words, arbitrary boundaries were (unconsciously) internalized by the Korean population in Japan.

This reassertion of sovereignty in a colonialist manner has significant implications for the geopolitics of the islands located between Japan and the Korean Peninsula. During and after the U.S. occupation, these islands became the main arena for the suppression of irregular migrants. In this sense, the islands functioned as border zones where the sovereignty of a nation-state was at work. However, numerous extralegal practices in Sasebo, the hinterland of these islands, show that this border did not serve a singular function. Their peripheral character, as perceived by both the ROK and Japan, was substantially different from the one that emerged in response to the Russian threat at the end of the 19th century. While the islands were included in a national territory that clearly belonged only to a particular state, their relations across the sea area formed under Japanese colonialism had to be excluded. Koreans who migrated at risk and were arrested did not merely violate the boundaries but were directly exposed to colonialist sovereignty enacted in the detention camp in Sasebo. Moreover, the dialectic of inclusion/exclusion that emerged in the relationship between the islands and Sasebo was applied to all Koreans residing in Japan.

Postwar Japanese society had to accommodate various migrations originating from its colonial past; in addition to Koreans, repatriates from former colonized regions such as Manchuria, or the South Seas were included. Asano (2010) describes this situation most aptly as a "folded-up empire," which, of course, also connotes the spatial contraction of the existing territories. Newly produced frontiers indicate that the geopolitical characteristics of the islands have largely been neglected within static center-periphery relationships, and their fluctuating nature and differential expression deserves a proper acknowledgement.

6 Conclusion

In 1951, the detention camp in Sasebo was closed and relocated to Omura, where naval air facilities were in operation (Fig. 2). After the US occupation, Koreans' legal status was eventually determined as the loss of their Japanese nationality, as stipulated by the San Francisco Peace Treaty in 1952. During the 1950s, negotiations between Japan and the ROK regarding the normalization of diplomatic relations were deadlocked. This situation once again exposed the fluctuant nature of the borders in this sea area (Hyung 2007). While the ROK government refused to accept the deportation of criminal offenders (not irregular migrants), Japanese fishermen were captured and detained in Busan for violating the so-called Peace Line established by Park Chung Hee, the President of the ROK. At that time, Omura and Busan faced each other across the straits. Detention camps in both cities were established as a leverage of sovereignty, where each state could not carry out deportations or fully exercise its own sovereignty.

With the normalization of diplomatic relations between Japan and the ROK in 1965, the boundary between the two countries was mutually recognized, except for the title of Takeshima/Dokdo. Consequently, the shipping route between Busan and Shimonoseki was reopened, making official traffic through the strait possible. As a result of ROK's economic development, irregular immigration to Japan decreased, and the islands became less of an issue in Japanese immigration policy. Meanwhile, once successfully landed, irregular migrants continued to face the threat of deportation. As their stay in Japan increased, some reported to the Immigration Bureau and gambled on the possibility of obtaining special permission to stay and regularize their residence, risking deportation. While the probability that their claim would be accepted was relatively high for those born in Japan, they had no choice but to be subjected to violent sovereignty. At the same time, Koreans legally residing in Japan were not only haunted by such sovereignty,[9] but also suffered from a lack of civil rights caused by the forthright and unilateral removal of their Japanese nationality. Araragi (2019) accurately described the situation of Koreans in postwar Japan as

[9] On the relationship between the 'haunting' of sovereignty and biopolitical practices, see Mountz (2011).

"domestic deportation." The memory of Sasebo and its associated islands has faded, but the colonialist sovereignty reconstituted through them lives on.[10]

Meanwhile, after the termination of the U.S. occupation of Sasebo, the U.S. naval base continued to be located there. It thus remains a place where U.S. sovereignty is still exercised. Sasebo is one of the main ports of the U.S. Seventh Fleet, which covers not only islands around western Japan but also those in the East and the South China Sea. Here is a good example of how 'spaces of exception' persist, while they can morph and change according to the specific requirements of the moment.

References

Agamben G (2005) State of exception. University of Chicago Press, Chicago

Araragi S (2019) Repatriation, deportation, and remaining: cases in the postwar East Asia. In: Araragi S, Kawakita A, Matsuura Y (eds) Repatriation, deportation, and remaining: comparative studies of international migration after the World War II. Nagoya University Press, Nagoya, pp 18–44 (in Japanese)

Asano T (2010) Postwar Japan as a "folded-up empire" and the formation of the East Asian region. J Jpn Intellect Hist 42:15–24 (in Japanese)

Chung Y (2012) Alien registration ordinance and Korean organization in Japan: Focusing on the process of registration. Bull Kyoto Hum Rights Res Inst 17:117–151 (in Japanese)

Division of Social Affairs, Ministry of Health Affairs (n.d.) Registration of Koreans, Chinese, Taiwanese, and persons whose family register is south of 30 degrees north latitude (including Kuchinoshima Island, Kagoshima Prefecture, or Okinawa Prefecture) conducted on March 18, 1946. Tokyo: Division of Social Affairs, Ministry of Health Affairs

Fukumoto T (2008) Certification of Korean "illegal entrants" and colonialism under U.S. occupation. In: Araragi S (ed), International sociology of migration evoked by the Japanese Empire, pp 147–175. Kyoto: Fuji Shuppan (in Japanese)

Fukumoto T (2013) The management of "illegal" Korean entrants during the U.S. occupation period and the persistence of colonialism: focusing on the detention camp in the Sasebo Repatriation Relief Bureau. In: Araragi S (ed) Human mobility after the Japanese 'Empire': intersection between post-colonialism and globalism, pp 477–510. Tokyo: Bensei Shuppan (in Japanese)

Hyakutake S (1951) Sad stories of illegal entrants. In: SRRB (ed) The history of the Sasebo repatriation relief bureau, vol 2, p 226. Tokyo: SRRB (in Japanese)

Hyung M (2007) Illegal entrants, Omura detention center, and Jeju Island: Networks of "illegal entrants" connecting Osaka and Jeju Island. Gendai Shiso 35(7):158–173 (in Japanese)

Ishibashi M (1950) What I desire to Hario. Keiko, May 1950, pp 37–39 (in Japanese)

Kagino H (1951) Daily reports. In: SRRB (ed) The history of the Sasebo repatriation relief bureau, vol 2, pp 228–229. Tokyo: SRRB (in Japanese)

Kim T (1997) Politics in the postwar Japan and the problem of Korean residents: policies of SCAP toward Koreans in Japan, pp 1945–1952. Tokyo: Keiso Shobo (in Japanese)

[10] Under the new residency management system enacted in 2012, which replaced the previous Alien Registration Law, foreigners are now listed in the Basic Resident Registration as "residents" in the same way as Japanese nationals. To be noted here is the treatment of foreigners that is considered "illegal." Under the new system, for example, overstayers are no longer allowed to register as residents, which was previously possible, and they are excluded from accessing even minimal public services.

Kobayashi S (2011) Politics of mobility of Koreans: Japan and southern Korea under U.S. occupation. In: Kishi T (ed) Self-portrait of modern Asia and others: local societies and problems related to "foreigners", pp 107–130. Kyoto: Kyoto University Press (in Japanese)
Kyoto Branch of Central Liaison Office (1948) Semimonthly report, Kyoto branch of central Liaison office, No.11, the latter half July. Kyoto: Kyoto Branch of Central Liaison Office (in Japanese)
Martin LL, Mitchelson ML (2009) Geographies of detention and imprisonment: interrogating spatial practices of confinement, discipline, law, and state power. Geogr Compass 3(1):459–477. https://doi.org/10.1111/j.1749-8198.2008.00196.x
Morita Y (1996) The numbers tell the story of Korean residents in Japan. Tokyo: Akashi Shoten (in Japanese)
Morris-Suzuki T (2010) Borderline Japan: foreigners and frontier controls in the post-war era. Cambridge University Press, Cambridge
Mountz A (2011) The enforcement archipelago: detention, haunting, and asylum on islands. Polit Geogr 30(3):118–128. https://doi.org/10.1016/j.polgeo.2011.01.005
Mountz A, Coddington K, Catania RT, Loyd JM (2012) Conceptualizing detention: mobility, containment, bordering, and exclusion. Prog Hum Geogr 37(4):522–541. https://doi.org/10.1177/0309132512460903
Nishioka S (1950) The past and present in Hario. Keiko, May 1950, pp 33–36 (in Japanese)
Otao T (1949) Episodes in repatriation reliefs. In: SRRB (ed) The history of the Sasebo repatriation relief bureau vol 1 (appendix pp 20–21). Tokyo: SRRB (in Japanese)
Reid-Henry R (2007) Exceptional sovereignty? Guantánamo Bay and the re-colonial present. Antipode 39(4):627–648. https://doi.org/10.1111/j.1467-8330.2007.00544.x
Ri Y (2023) Social history of immigration control: the management of "boundaries" in the postwar Japan. Tokyo: Akashi Shoten (in Japanese)
Seki F (1950) Tediousness in Hario. Keiko, May 1950, pp 43–45 (in Japanese)
Security Division, Nagasaki Prefectural Police (1950) Overview of coastal security in the prefecture. Keiko, May 1950, pp 28–32 (in Japanese)
SRRB (Sasebo Repatriation Relief Bureau) (ed) (1949) The history of the Sasebo repatriation relief bureau, vol 1. Tokyo: SRRB (in Japanese)
SRRB (Sasebo Repatriation Relief Bureau) (ed) (1951) The history of the Sasebo repatriation relief bureau, vol 2. Tokyo: SRRB (in Japanese)
Suzuki K (2006) Transition of assistance policies for repatriation of Koreans in Japan: the case of Shimonoseki and Senzaki. Zainichi Chosenjin Kenkyu 36:163–184 (in Japanese)
Takahashi S (2011) Governing "others" and citizenship: a focus on the mechanism of inclusion/exclusion and its emergence in grey zone. Shakai-Shirin Hosei J Sociol Soc Sci 57(4):253–268 (in Japanese)
Wagner WW (1954) The Korean minority in Japan, 1904–1950. International Secretariat, Institute of Pacific Relations, New York
Yoshimura K (1963a) Hario Sub-blanch heavily busy with deportation: an aspect of the immediate aftermath of the war from the police's view, part 2. Keiko, May 1963, pp 12–16 (in Japanese)
Yoshimura K (1963b) Hario Sub-blanch heavily busy with deportation: an aspect of the immediate aftermath of the war from the police's view, part 3. Keiko, August 1963, pp 38–42 (in Japanese)
Yoshimura K (1963c) Hario Sub-blanch heavily busy with deportation: an aspect of the immediate aftermath of the war from the police's view, part 3. Keiko, August 1963, pp 50–54 (in Japanese)

Developing Border Tourism for the Sustainability of Border Island: The Case of Tsushima Island in Japan

Yasunori Hanamatsu

Abstract This chapter discusses the role that border tourism plays in enhancing the sustainability of a border island by focusing on Tsushima Island in Nagasaki Prefecture. Located between Kyushu, Japan and Busan, South Korea, Tsushima has played an important role as a front line both of national defense and international exchange between Japan and the Korean Peninsula since ancient times. There have been no territorial disputes except for a period in the Middle Ages, and in intergovernmental relations sovereignty is clearly delineated, making it the most legally stable borderland in Japan. Therefore, the traffic of people is active, and in 2019, more than 400,000 foreign tourists visited Tsushima annually. Most of them are Koreans, but for the sustainability of Tsushima, which is located in the border region of Japan territory and has a declining population and industry, the acceptance of Korean tourists is one of the lifelines. On the other hands, Japanese tourists do not usually cross the border between Tsushima and Busan and it is expected to develop such a cross-border tourism by Japanese tourists. Recently, efforts to enhance sustainability by developing border tourism have faced serious challenges; deterioration of the Japan-South Korea diplomatic relations, the COVID-19 pandemic and a temptation to receive subsidies from national government by accepting nuclear waste. Based on the results of fieldwork and interviews conducted in Tsushima for more than 10 years, this chapter examines the possibility and significance of border tourism for enhancing the sustainability of Tsushima island.

Keywords Tsushima · Border island · Border tourism · Busan · Territorial trap

Y. Hanamatsu (✉)
Faculty of Law, Kyushu International University, Kitakyushu, Japan
e-mail: hanamatsu@law.kiu.ac.jp

T. Yamazaki and G. Baldacchino (eds.), *Relational Geographies of Islands in the Indo-Pacific*, International Perspectives in Geography 23,
https://doi.org/10.1007/978-981-96-7155-7_7

1 Introduction

This chapter discusses the role that border tourism plays in enhancing the sustainability of a border island, focusing on Tsushima (Tsushima City, Nagasaki Prefecture), which is a typical example of a remote border island in Japan (Hanamatsu 2022).

Japan is an island nation surrounded by the Pacific Ocean, the Sea of Japan, and the East China Sea, and consists of the four relatively large islands (or mainland) of Honshu, Hokkaido, Shikoku, and Kyushu, as well as more than 14,000 islands. All of Japan's borders are at sea, and islands near these borders function as "border islands."

For many people living in Japan, the three border-disputed areas—the Senkaku Islands, Takeshima, and the Northern Territories—capture the public imagination (Hosaka and Togo 2012). Borders are otherwise rarely considered in most people's daily lives. Under such circumstances, the news of these conflict areas that are occasionally reported in newspapers and on television tend to form the general view and establish the narrative about the borders of Japan. In other words, the image of a border is that it is dangerous, inaccessible, hostile and a "dead end" or "endpoint": it is a view of the border as a hard boundary between sovereign states that are monolithic and powerful entities that confront each other spatially (Iwashita 2016). Such a view is akin to the concept advocated by classical geopolitics during the imperialist era (Yamazaki 2017; Kawakubo 2018). In recent years, since tensions between Japan and neighboring countries such as China, South Korea, and Russia tend to be emphasized, it is not surprising that such a classical understanding of borders tends to be accepted uncritically. Nevertheless, there are other border regions in Japan that such a dark, classical view of borders cannot explain. This is the case of Tsushima island.

Tsushima has sought to maintain its socio-economic sustainability (Thompson and Norris 2021) by taking advantage of its remote border status and developing cross-border ties with neighboring South Korea. Breaking away from the perception of borders or borderlands as "dead ends," "frontiers," or "fortresses," Tsushima has promoted a unique perception of borders as "gateways" (Iwashita and Hanamatsu 2014) and "meeting points" (Mogiani 2023). On the other hand, there is also a dynamism that tries to balance national and global movements that push borders back into the perception and practice of "dead ends" and "frontiers." This chapter introduces the dynamics of efforts by one of Japan's border islands to increase its sustainability by developing border tourism.

2 Tsushima Island

Tsushima is a border island located in the northwesternmost part of Japan, located between the Tsushima Strait and the Korea Strait. As an administrative district, it belongs to Nagasaki Prefecture and constitutes a single municipality: Tsushima City (Fig. 1). With a population of about 28,000 in 2023, it is the 10th largest island by land area in Japan. About 90% of its surface is covered with mountains and forests, and there is little land suitable for cultivation. Thus, since ancient times, people have lived through exchange and trade with the Korean Peninsula, which is 49.5 km away (Fig. 2). Before the establishment of the modern concept of territory, there was a period of ambiguity in the island's political affiliation (Saeki 2014). However, in recent centuries, the fact that Tsushima is a Japanese territory is not disputed between the neighboring states, and the border drawn between Tsushima (Japan) and Busan (South Korea) has been fully agreed upon between the respective national governments. It is rare in Japan to have a border region where border lines are demarcated in this way. In most border regions other than Tsushima, specifically those adjacent to Russia, China/Taiwan, and other parts of South Korea, no clear border line is agreed upon, or there are de facto territorial disputes. And because the border in this area is clearly demarcated, Tsushima has accepted many tourists from Busan for about 20 years, on the opposite side of the strait, and formed a particular perception of the border that boosts its habitability and sustainability.

In Tsushima, sustainability is often discussed from the perspective of the environment and ecosystems. More than 130 km from mainland Japan and only 50 km from the continent, Tsushima has a unique ecosystem with a mix of endemic flora

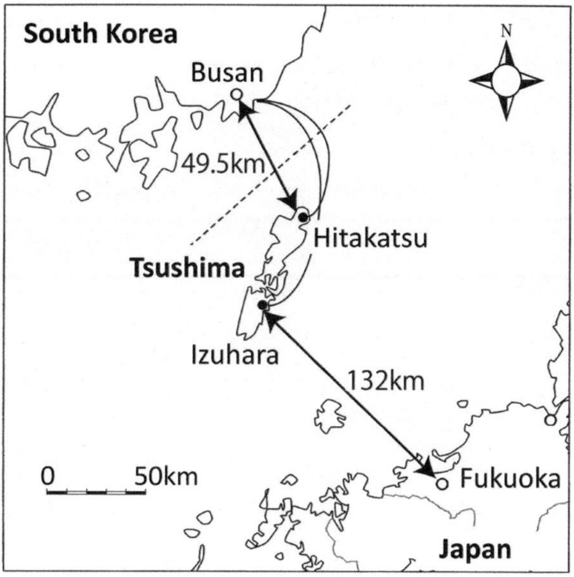

Fig. 1 Location of Tsushima Island

Fig. 2 Cityscape of Busan from Tsushima Island (photo by the author on 8 October 2017)

and fauna influenced by both the Japan archipelago and the Eurasian continent. A typical example is the endemic Tsushima leopard cat (*Prionailurus bengalensis*), a Far Eastern subspecies of the Bengal wildcat, threatened by habitat loss. On the other hand, a large amount of marine waste flows into Tsushima on the currents flowing from the East China Sea into the Japan Sea, threatening the ecological sustainability of the island. Additionally, in 2023, Tsushima has recently been mooted as a disposal site for high-concentration radioactive waste generated by nuclear power plants in Japan (Asahi Shimbun 2023). In this way, the sustainability of the environment and ecosystems is a popular conversation item in Tsushima; this chapter, however, dwells on the sustainability of Tsushima from a socio-economic perspective.

To support sustainable livelihoods on the island, suitable, reliable and affordable transportation to and from the mainland and urban areas of the country is essential. The nearest metropolitan areas, Fukuoka City in Kyushu and Nagasaki City, the prefectural capital, are served by about ten round-trip flights a day. There is a ferry service connecting Fukuoka, about 130 km away, with about five daily round trips. Although it can be affected by weather conditions, transportation to and from metropolitan areas is considered to be largely secure. Yet, not many Japanese tourists come to Tsushima from within Japan, and there are only about 100,000 people a year, including business travelers (Tsushima City 2022). It is estimated that only tens of thousands of Japanese tourists come to Tsushima for tourism annually. The tourism industry targeting the Japanese is not thriving well.

On the other hand, because it is a remote border island, the entire island of Tsushima has functioned as a military fortress since the late nineteenth century, and the development of roads and other facilities on the island has been suppressed for many years. As folklorist Tsuneichi Miyamoto once said, "Tsushima still has its Middle Ages" (Miyamoto 1984); Tsushima was one of the most "backward" regions of Japan. Since modern times, fishing has become a key industry and has been booming, but it has suffered with the decline of marine resources in the seas around Japan, and the population has fallen to less than half from the 1960s, when it was at its peak.

Along with the declining population, the declining birthrate and aging population are serious concerns. From 2013 to 2017, Tsushima City's total fertility rate was 2.21, the 11th highest among municipalities in Japan (Japan Ministry of Health 2020). However, due to the decline in the population of the parent's generation, the number of births and the total number of children continue to decline steadily; and, in recent years, elementary and junior high schools have been forced to close almost every year. Many children leave the island to go on to higher education or find employment upon graduating from high school, and most of them do not return to the island. According to estimates by the National Institute of Population and Social Security Research, Tsushima's population will decline to about 13,000 by 2045 (Tsushima City 2020). On the other hand, the aging rate, which indicates the proportion of people aged 65 and over in the total population, was 37.9% in 2021 for Tsushima City as a whole (Tsushima City 2021): this figure is estimated to be similar to the average aging rate of Japan in 2050 (Cabinet Office, Government of Japan 2022). In addition, in some areas of the island, the aging rate has already exceeded 60%. Population decline, declining birthrate and aging population are problems for Japan as a whole (Konishi et al. 2018); but in Tsushima, the number of local people who support the elders is minimal due to the severe declining birthrate.

Under these circumstances, in 2016, the Japanese government enacted the "Act on Preserving Remote Island Areas" (The Act on Preservation of Areas of Remote, Inhabited Islands Establishing Territorial Seas and Maintenance of Local Societies on Areas of Specified Remote, Inhabited Islands Establishing Territorial Seas). This law designates remote border islands that serve as bases for preserving territorial waters and exclusive economic zones (EEZs) as "areas of remote, inhabited islands establishing territorial seas" and works to prevent uninhabited islands. Particularly, in the "an area of a specified, remote, inhabited island establishing territorial seas" far from the mainland and where the population is declining significantly, the Japanese government provides financial support for population maintenance measures. Tsushima has been designated as one such area, and has therefore received state subsidies to lower shipping costs and airfares for residents, expanding employment opportunities by supporting business start-ups and business expansion, reducing the burden of transportation costs for goods, and promoting long-stay tourism (Furukawa 2020). However, although some results have been achieved in the increase in the number of immigrants from outside the island and the launch of new businesses, the current situation is that the population decline and declining birthrate in Tsushima have not been stopped, let alone reversed.

In this way, the stagnation of key industries, the declining population, the declining birthrate, and the aging population are progressing, and the sustainability of Tsushima as a typical remote border island in Japan is at risk. On the other hand, Tsushima is trying to improve the island's sustainability through cross-border exchanges with Busan in South Korea on the opposite coast.

To accurately understand the implications of this dynamic process, one must consider the following three analytical frameworks.

3 Analytical Frameworks

3.1 The Territorial Trap

As mentioned above, the border region of Japan consists of the sea and islands, so for many Japanese, the border region can be captured as an image of "frontier" or "dead end." As a remote border island, Tsushima is viewed as a "frontier" island far from urban areas; hence, it is likely to be imagined as inevitably challenged by a most disadvantageous socio-economic situation. However, as described later, Tsushima has been trying to overcome its disadvantageous situation through cross-border exchanges with Busan. Why is this kind of initiative by Tsushima not properly understood by many Japanese people?

One of the cognitive frameworks that makes it difficult to grasp the actual situation of borders and border regions is the "territorial trap" conceptualized by the American political geographer John Agnew (Agnew 1994). This is not the trap of territory itself but the trap of territoriality, that is, an epistemological trap into which we tend to fall when thinking about territory (Yamazaki 2014).

According to Agnew (1994), the debates developed in international politics and international relations are generally caught in "the territorial trap." The trap specifically consists of the following three elements. The first is the idea that a state should be able to exercise exclusive sovereignty over a clearly demarcated territory; the second is the recognition that there is a clear distinction between within and outside the state, domestic and foreign affairs, domestic and international politics across the border; and the third is a recognition that the boundaries of the state and the boundaries of society should align and coincide.

Based on this recognition, the exclusive control of the territory will be established by clearly drawing borders, domestic affairs and foreign affairs will be clearly separated, and the border regions will be "frontiers" or "dead ends". The three border dispute areas in Japan introduced at the beginning of this chapter are often discussed based on this perception. However, the "territorial trap" approach does not allow us to grasp the actual situation of the border region between Tsushima and Busan, as well as many border regions around the world. The situation has temporarily worsened due to the deterioration of Japan-Korea relations and the COVID-19 pandemic; but exchanges with Busan, on the other side of the strait, remain significant for Tsushima.

Nevertheless, if we consider Tsushima as a single island unit with (in Japanese) territory, we will not be able to appreciate any internal differences and discontinuities in border regions. To accurately grasp these realities, another analytical framework is required, as described below.

3.2 Border Studies

Border Studies provides an epistemic framework that escapes from the "territorial trap" that considers national territories in a uniform or fixed manner and regards borders as the limit line to which the power of sovereign states extends. It is a relatively new academic discipline that has tried to explore and theorize the diverse dynamics and meanings of borders and border regions (Iwashita 2016).

Based on international law and political geography, which consider the location and effects of "national borders" as typical examples of spatiality and physicality, border studies try to target all the borders that exist in the world, questioning the function of delimited space, as well as the meaning of delimiting practices itself. These studies emphasize the concept of "borders as socially constructed" while incorporating the knowledge of critical geopolitics. In other words, it is understood that borders are not mere physical or material artefacts, but constructed by and out of human imaginations and their actions of "bordering" (Newman 2006). From this perspective, it is possible to examine "who drew the border and how?" and "what constitutes the border?" (Kawakubo 2018).

Based on such re-conceptualization of borders, border studies make three important contributions. First, a border is reinterpreted not as a boundary, but as a spatial "zone" that straddles the border line: this is called a border region or borderland. In addition, it proffers a new approach to grasp the dynamic process of changes which are unique to such borderlands. This approach makes it theoretically possible to grasp the actual situation of the border region straddling Tsushima and Busan, which would not be seen from the epistemological strictures of the "territorial trap."

Second, borders are understood as a dynamic process of "de-bordering," "re-bordering," and "neo-bordering," in which a border is not fixed or static, but disappears, is reconstituted, and/or is newly created (Rumford 2006). As long as the construction of borders depends on human cognition and behavior, the borders themselves and their meanings can constantly change both spatially and cognitively. For example, human collective perception may determine the cognitive interpretation of physical borders such as national borders (Diener and Hagen 2012).

Third, one can move away from the state-centric view of borders, which sees borders only as a place for the exercise of power by the state or as a symbol of territorial control, and to consider how those living in the border regions and the societies formed there face, navigate and negotiate the borders in their daily lives; and how their behaviour and actions affect the dynamics of these borders, for example, rendering them more or less porous. Borders can exist anywhere, and "borderwork" is not the exclusive property of the state but is open to anyone (Cooper et al. 2014).

By focusing on border societies and their inhabitants who cope with borders in their daily lives, it becomes clearer that border conceptualisations have been arbitrarily reproduced by a rigid and unidimensional framework implied by nation-states and international relations.

3.3 *Border Tourism*

One of the phenomena that can be observed by the epistemological framework provided by border studies is border tourism. This is a way of traveling that focuses on all types of borders, including national borders, adjacent border regions and borderlands.

The purpose of border tourism is to identify the border as a destination per se, not just as a transit point on a journey, and to experience the border itself, as well as its peculiar landscapes, people's lives, politics, geography, economy, culture, and history. The term is usually used to refer to travel to border regions that exist between countries, although it also covers borders between municipalities or regions and boundaries with areas that are not easily accessible (military bases, nuclear power plants, etc.) within national territories.

Borders are often perceived as dividing lines that control the movement of people and goods (Timothy 2001). It can also be assumed that, since borders are barriers to tourism, the two are incompatible. As mentioned earlier, the dangerous and dark images of border such as "dead ends," "frontiers," and "end points" evoked by border conflicts tend to keep people away from borders. Border regions as "front lines of national defense," "fortresses," and "barriers" are generally not considered conducive to become tourist destinations, even if they are subject to national security (Iwashita 2016). On the other hand, border regions are also the "front line of cross-border exchange." In these regions, there is a unique culture and history that develops while mixing in close relationships with neighboring countries, and there are activities of people who confront and make use of national borders in their daily lives. From this perspective, border regions come across as "gateways" (Iwashita 2016), and borders can be regarded as "meeting points" or "linking points" that connect countries (Mogiani 2023). This ambivalence is a major feature of the border regions in terms of geopolitics, and it is an attraction for tourists that cannot be found anywhere else (Hanamatsu 2016).

Border tourism can be classified into the following four categories according to its objectives (Hanamatsu 2020). The first is interest in border lines and border regions themselves. For example, approaching the border regions of walls, fences, and barbed wire satisfies curiosity about a unique space and the military tensions in which the border artificially divides space and restricts traffic. The second is the act of border-crossing. In today's world, where it is possible to travel freely by air without being aware of national borders, the act of border-crossing by land or sea is an extraordinary experience in itself, and which reaffirms state-constructed frontiers. The third is to enjoy the differences and benefits created by the existence of borders. In addition to duty-free shopping, opportunities for cross-border shopping allows consumers to take advantage of differences in price, culture, and quality. Medical and gasoline tourism would fall within this category. In addition, one of the objectives will be to experience a different culture easily, inexpensively, and quickly. The fourth is to learn about the history of changes in national borders and the fusion of cultures across borders. Borders are not immutable, and the process of de-bordering, re-bordering, and neo-bordering is historically contingent. It is a learning journey that follows this historical process and explores how borders that existed in the past can still influence current relations between nations.

One can assume that the Korean border tourists who come to Tsushima are mainly motivated by the third category of purpose: they can enjoy the differentiation made by borders, as described in more detail below. If the main function of geographical borders is to delimit spaces and to establish and highlight differences between them (Diener and Hagen 2012), then a purpose of border tourism is to enable an experience of such differences away from one's own daily space and familiar territory. In this sense, tourism is inherently closely related to borders. The border itself differentiates the space and the politics, economy and culture associated with it. Experiencing such differences is an incentive for tourism (Urry and Larsen 2011). In that case, an examination of what kind of differentiation is made by the border drawn between Tsushima and Busan, and how this promotes border tourism, is warranted. Meanwhile, very few Japanese border tourists cross the border the other way: from Tsushima to Busan. Potential reasons for this will be examined later.

The author has been conducting fieldwork research in Tsushima for more than 10 years. Over a long period of time, he has repeatedly visited Tsushima's local government officials, mayor, local assemblymen, tourism operators, fishermen, ship operators, and local NPOs, and established a relationship of trust over a long period of time. In addition, he has interacted with residents and local high school students during the fieldwork in Tsushima. Through interviews, he has gathered information about how local people in Tsushima have viewed the border, neighboring countries, and Korean tourists. On the other hand, he has also conducted interviews and questionnaire surveys of Korean and Japan tourists, for clarifying why they visited Tsushima. Based on the results of previous field surveys, the next section examines the possibility and significance of border tourism for enhancing the sustainability of Tsushima Island.

4 Border Tourism in Tsushima

4.1 Borders as a "Resource"

When considering border tourism, it is important to consider both the perspectives of travellers, as well as the implications for the border region. For border regions that are economically and socially disadvantaged as "frontier" regions, the border could be a valid "resource" for tourism (Sohn 2014).

Tsushima has struggled to develop the region by changing the image of a remote border island as a "dead end" by exploiting its relative geographical proximity to Busan as a resource. The maritime boundary between Busan and Tsushima is fixed, and there are no physical barriers such as walls or fences, making it the most stable border region in Japan. Historically, Tsushima had a close relationship with the Korean Peninsula. But, even after diplomatic relations were restored with the 1965 Basic Treaty between Japan and South Korea and the border with Busan was newly established, there was still very little civilian traffic between Tsushima and Busan (Nagadome 2009). In the meantime, while the local population and industries, such as fishing and forestry, gradually declined, new efforts were launched to maintain the island's sustainability by promoting cross-border exchanges with Busan, a mega city of 4 million people on the island's opposite shore. Since the 1980s, when there were no international shipping routes with Busan, the Tsushima local government has been conducting exchange projects to cross the strait by kayak and long-distance swimming. In addition, local development programs sensitive to the historical relationship with South Korea have been actively implemented. These include holding events celebrating "Chosen Tsushinshi" (Korean Diplomatic Delegations to Japan during the Edo period) that re-established diplomacy and peaceful relations between the two countries after Japan invaded Korea in the late sixteenth century, and where the Tsushima Clan played a vital role of coordinating the missions.

The turning point was the opening of international shipping routes. In 1989, using subsidies from the Japanese government, Tsushima launched a ferry service between Tsushima and Busan, operated by small boats. Subsequently, in 1999, a route connecting the port of Busan and the port of Izuhara, the island's central city, was financed by Korean capital, and the following year, it became a regular route. In 2001, a shipping route connecting Busan Port and Hitakatsu Port in the island's northern part was launched, quickly becoming a daily operation (Fig. 3). According to the "Immigration Control Statistics Table" disclosed by the Immigration Services Agency of the Government of Japan, as a result, the number of tourists from South Korea has increased yearly, from 8,000 in 2001 to 42,000 in 2006, and began to grow rapidly in 2012 when three companies started to operate about ten round-trip services every weekend. The number of arrivals in 2015 exceeded 215,000 and 410,000 in 2018. About 98% of tourists are Koreans (Immigration Services Agency, Government of Japan). Even though the number of inbound visitors to Japan was increasing, it was rare that on an island with a population of only 30,000, about 13 times as many Korean tourists landed. While domestic tourism and business visitors

Fig. 3 International passenger Terminal at Hitakatsu port (photo by the author on 27 August 2017)

from Japan remained almost flat, tourism consumption on the island grew rapidly, and the island's economy boomed. According to the Tsushima local government, in 2018, 410,000 Korean tourists spent about 9.1 billion yen (US$63 million), about one-quarter of the annual budget of Tsushima City in the same year.

Moreover, in the wake of the rapid increase in the number of Korean tourists, a new project was carried out to improve maritime connectivity between Tsushima and the metropolis of Fukuoka, on Kyushu. A ferry service between Fukuoka, Tsushima (Hitakatsu) and Busan was launched in 2018, combining domestic and international travel: a first for Japan (Furukawa 2020; Takagi 2023). Domestic passengers can board between Tsushima and Fukuoka without carrying their passports on the international ferry service between Fukuoka and Busan. Hitakatsu Port in northern Tsushima is a gateway for many Korean tourists to enter the country, but until then, there was no domestic high-speed ferry service connecting Hitakatsu to Fukuoka. For the older people who live in northern Tsushima, which is quite far from Tsushima Airport, it was a long-cherished desire to have a means of transportation that would allow them to easily commute to hospitals in Fukuoka, where medical services are satisfactory. Such a transportation initiative is only possible because of the development of border tourism fueled by Korean tourists between Tsushima (Hitakatsu) and Busan, enhancing Tsushima's sustainability by using the border as an economic resource. Regretfully, the international route between Fukuoka, Tsushima and Busan was suspended due to COVID-19 in March 2020, and there is still no prospect of revival, even though the international route between Tsushima (Hitakatsu) and Busan has resumed in 2023.

4.2 Familiarity and Difference in a Borderland

Why do so many Koreans visit Tsushima? In the past, the purpose of a tour of Tsushima by a group of Koreans was duty-free shopping at the passenger terminal of the port of Busan. The increase of Korean border tourists in recent years has been driven by cheaper high-speed ferry tickets, geographical proximity, the permanent visa waiver in 2006, and the recent depreciation of the Japanese yen. Other significant explanatory factors include the "familiarity" and "difference" recognized across national borders.

The author has conducted long-term interview surveys in various parts of Tsushima, targeting Korean tourists, travel agents, local government, and residents since 2013. According to the interviewees, the rich natural habitat of Tsushima, left untouched as a border island, has won the hearts of many Koreans (Hanamatsu 2016). Trekking, cycling, and fishing are popular among them. Walking through the streets of Tsushima appeals to Koreans as a "familiar but somehow different and unlikely" exotic experience. There is also a specific "borderscape" (Brambilla et al. 2015). Signboards displayed in Korean are everywhere in the city. Although Busan is 50 km away, one can still receive a mobile phone signal from Korean. This kind of "scenery and experience that is Japan but not Japan" gives Koreans with a unique sense of wonder (Hanamatsu 2016).

Korean shopper now targets Japanese snacks, seasonings and alcoholic beverages at local supermarkets as well as shopping at duty-free shops on the island. During the fieldwork in Hitakatsu from 2013 to 2018, Koreans have often been seen buying large quantities and packing purchases in suitcases or cardboard boxes before returning home. Since it is possible to make a round day trip for a few thousand yen, it may be an image of "extending our journey a little and shopping there" rather than sightseeing fully prepared for Korean tourists. Koreans regard Tsushima as the "closest and cheapest Japan." Although it is a sightseeing tour of Tsushima, experiencing Japanese culture, history, and nature is becoming a major purpose (Iwashita and Hanamatsu 2014).

Enjoy the familiar but somehow different is probably the key charm of Tsushima for Koreans. This is close to the concept of "expected unfamiliarity" proposed in border studies (Spierings and Van der Velde 2008). People naturally want to enjoy experiences that are different from their own familiar places, cultures, languages, landscape, goods, and economies. The unfamiliarity of places can be seen as an incentive to cross borders because they promise "new" experiences (Timothy 2001). However, large differences in the socio-cultural sense can result into mental borders that have a negative impact on cross-border interaction (Van Houtum 1999). When differences between places on both sides of the border are too large, this could restrain people from visiting these places (Spierings and Van der Velde 2008). The "differences" and "unknowns" formed by the borders must not be overly different but rather need to be located within a foreseeable range that is easy to accept. In other words, an environment in which familiarity and difference coexist in an appropriate balance is vital to motivate people to move across borders (Spierings and Van der Velde 2013). Thus, the border that produces "expected unfamiliarity" is a boon for tourism: this is the case for the border island of Tsushima. The extent to which we can shift from a view of borders as a negative condition that militates against the movement of people and goods, to a view of borders as a resource that promotes cross-border movement by providing an appropriate difference in space through barriers is quite likely to determine the future sustainability of Tsushima.

4.3 Transformations in Tsushima Through Border Tourism

Tsushima society accepts many Korean tourists, and has also come to embrace the various discontinuities between the two cultures. By conducting fieldwork and interviews in various parts of Tsushima, the author has seen the challenges and tensions of the people of Tsushima in the face of Korean tourists, who consist mainly of older travellers and can be responsible for some behavioural issues. There have been reports of Korean tourists opening products at supermarkets before checkout and bringing kimchi (Korean pickles) from home to restaurants and bars in Tsushima. Such episodes reflect cultural differences between the two neighbouring countries. In recent years, such events of cultural collision have decreased; Korean tourists are now younger, and the Tsushima local government has taken active measures to communicate Japanese manners to Korean visitors.

From the perspective of increasing revenue from Korean tourism and revitalizing an island that is suffering from population decline, there is a tendency for many islanders to reach out to Koreans, closing the culture gap. Even if they don't understand the Korean language, locals know that interacting with tourists is likely to build customer satisfaction and increase tourism revenue. In this sense, it can be said that the "de-bordering" process is developing, in which the cultural gap between residents in Tsushima and Korean tourists is narrowing.

This, however, is not to say that it has disappeared; some problems of cultural incompatibility remain (Hanamatsu 2016; Hirano 2000). The more they interact with each other, and the deeper they communicate with each other in daily lives, the more sensitive they may become to each other's differences. There are some restaurants and stores that explicitly declare "Koreans are not allowed," even though the exact reasons for this may not be explicit. Thus, "re-bordering" is progressing in parallel with the "de-bordering" process (Rumford 2006). During the fieldwork in Hitakatsu in 2017, the author found a signboard at the entrance to a restaurant saying "Fully Booked Today" written in Korean. But, when the author entered without a reservation, he was still allowed to sit without any problems. Is this because he was Japanese? As most of the local people in Tsushima do not understand Korean, they also usually enter a restaurant or store without noticing signboards written in Korean and are usually allowed to come in. In Tsushima, anyone can see this love-hate relationship, the dilemma of "exclusion" and "inclusion" toward Korean tourists by local residents, who practice "bordering" in a way that does not annoy Koreans while controlling their likes and dislikes and avoiding open social conflicts with Koreans.

According to the author's interviews with a local NPO conducted in 2022 and 2023, in recent years, there have been civic campaigns on the island in favour of legislating guidelines on the installation of signboards in the Korean language. The locals argue that it does not matter if the signboard of the shop contains Korean; but the main language on any billboard should always be Japanese, and that this should be stipulated at law. They also argue that shrines are traditional religious facilities in Japan and have their own rules of worship, so tourists must also follow these rules regardless of nationality. These can be seen as an example of "neo-bordering" by drawing new borders based on mutual consultation and rule-setting for mutual co-existence. In this way, "de-bordering," "re-bordering," and "neo-bordering" coexist in various areas and scales on the island.

4.4 Is an Island a Uniform Space?

The dynamism of a remote border island in Tsushima also needs to be understood in terms of the diversity that gradually develops in the residents of the island in terms of the perception of borders and their attitude toward accepting Korean tourists. Since islands exist as a limited geographical space surrounded by water, they tend to be regarded as homogenous and uniform. Nothing could be further from the truth. When visiting various areas of Tsushima Island during the author's fieldwork, the characteristics of each area on the island become clearer. Even though it is an island, Tsushima is not uniform: it is a large island that stretches about 100 km from north to south, with over 180 districts on the island, most of which are small villages consisting of a few dozen households. About 90% of the island is covered with forest and mountains, and small villages are isolated in the mountains. Local people in such small inland villages usually have little to do with Korean tourists, unlike Izuhara, the island's capital, and Hitakatsu, the gateway to South Korea. For local people in small villages, there is little awareness that Tsushima is a border island and has actually accepted many Korean tourists. Meanwhile, there are many residents in the fishing village on the west side of the island who are antipathetic to Koreans. During the author's fieldwork conducted since 2013, many residents on the island often told me that smuggling and poaching from the Korean side after World War II gave them bad vibes about the Korean Peninsula.

A gap has also gradually been clearer in the view of the border and Korean tourists between the two port towns that have been directly accepting Korean tourists for more than twenty years. Izuhara, the island center where the municipal office is located, is a historical town originating from a traditional castle town built by the So Family (a local clan in Tsushima) and has stronger ties with Fukuoka and Nagasaki on the mainland through both business and public services (such as government offices, hospitals and schools) than Busan. According to the author's interviews conducted since 2013, the author sometimes felt that they had a somewhat cold and despondent view of South Korea, even though they usually interaction heavily with Korean tourists. On one hand, Izuhara is characterized by civic activities to protect its identity as a "traditional castle town of Japan." For instance, city residents expressed a sense of alarm against installing store signboards written only in Korean by establishing certain rules for foreign language signboards and planning a project to display the Japanese national flag in the city.

On the other hand, Hitakatsu, the gateway to South Korea, is characterized by its open atmosphere and quiet and rich natural environment. During the author's fieldwork conducted since 2013, local government officials and tourism operators in Tsushima often told me that Korean tourists were more attracted to the distinguishing features of the northern part of Tsushima. Before the island's six small towns were consolidated into the larger Tsushima City in 2004, the town of Kami-Tsushima, located at the northernmost tip of the island, independently sought to promote mutual exchange with South Korea. Since the 1980s, an exchange project has been maintained in which participants cross over to Busan by kayaking and long-distance swimming. In 1989, the town launched an irregular ferry service between Hitakatsu and Busan by a small boat, using subsidies from the Japanese government. This northern area also has the geographical advantage of being only one hour from Busan by high-speed ferry and has actively accepted Korean tourists so far. At its peak, the area accepted about 70% of Koreans coming to Tsushima from Busan. However, it also had to struggle with "over-tourism" afterward.

In this way, even in port towns on the same island, there is a difference in their attitude toward accepting Korean tourists and their views of neighboring countries and borders. Nonetheless, if we see the island as a uniform space, we will fall into the territorial trap warned by Agnew (Iwashita 2016).

4.5 Is Border Tourism from Japan Possible?

Tsushima has developed border tourism that uses borders as "resources" by accepting many Korean tourists for more than twenty years. However, border tourism is not necessarily limited to inbound tourism by foreigners. Is it possible to promote border tourism in which Japanese tourists cross the border between Tsushima and Busan? To estimate this possibility, this section shows the results of experimental cross-border tours in collaboration with a major Japanese travel agency and questionnaires and interviews with tour participants.

Even before the deterioration of Japan-Korea diplomatic relations in 2019 and the disruption of tourists from Busan due to COVID-19, islanders shared a sense of crisis regarding the island's significant dependence on Korean tourists. The possibility that some accidents would affect tourists from Busan was widely understood. For example, after the Fukushima nuclear accident in 2011 and the sinking of the Sewol ferry in South Korea in 2014, the number of incoming Korean tourists decreased markedly. Therefore, while many islanders accepted Korean cross-border tourists daily, they wanted more Japanese tourists to visit their island. Nevertheless, few effective attempts were made to attract such domestic tourists.

Therefore, in order to propose a new type of travel that combines domestic sightseeing in Tsushima and overseas travel in Busan while incorporating the unusual experience of border-crossing by high-speed ferry, our research group piloted joint border tours three times—in 2013, 2015 and 2017—under the auspices of a major travel agency. Feedback from the participants was then collected (Hanamatsu 2017).

The participants positively accepted the novel concept of border tourism, which they had never seen before; neither has there been such a package tour in Japan. They usually sell domestic and international package tours separately, implying that they may be caught in the "territorial trap."

As part of the experimental cross-border tours, the participants and our research group visited sightseeing spots on Tsushima Island that show historical connections with the Korean Peninsula and crossed the border by high-speed ferry. Later, in Busan, we visited the site of the "Choryo-Wakan," where the Tsushima Clan stayed and traded with the Korean dynasty during the Edo period and experienced "Tsushima in Busan" which means the historical mutual exchange between Tsushima and Busan. Finally, we visited the museum of "Chosen Tsushinshi" (Korean Diplomatic Delegations to Japan during the Edo period) to learn the historical narratives from the Korean perspective, which is completely different from those of the Tsushima side.

During the second experimental cross-border tour from Fukuoka, Japan, to Busan, South Korea, via Tsushima in March 2015, the author conducted a questionnaire survey targeting twenty participants. According to the survey results, 90% of the participants were satisfied with the trial tour, and 75% wanted to revisit Tsushima and Busan combined (see Table 1). In addition, five out of six participants who had once visited Tsushima and Busan on different occasions supported border tourism between Tsushima and Busan. This indicates a special attraction to visit both regions across the border simultaneously. In addition, there was a comment that "I realized that the perspective of borders was missing." Therefore, it was found that such a border tour would provide the opportunity to promote some "awareness" and "learning" on borders, which most Japanese had forgotten about since Japan has no visible borders. Certainly, there are drawbacks, such as high prices because border package tours combining domestic and international travel are not simple round trips, and it is prone to emergencies such as diplomatic conflicts with neighbouring countries and infectious diseases. Nevertheless, Japanese tourists' cross-border tourism to Busan through Tsushima can contribute to the sustainable development of Tsushima and the improvement of the Japanese view of borders.

4.6 Challenges to Border Tourism

Recently, efforts to enhance sustainability by developing border tourism have faced serious challenges.

Table 1 Tour participants' responses to the selected questions (N = 20)

Q: Are you satisfied with this trial tour from Fukuoka to Busan via Tsushima?				
Highly satisfied	Satisfied	Neither satisfied nor dissatisfed	Dissatisfied	Highly dissatisfied
11	7	1	1	0

Q: Do you want to revisit Tsushima and Busan combined?			
I want torevisit Tsushima and Busan combined	I want torevisit only Tsushima	I want to revisit only Busan	I want to revisit neither Tsushima nor Busan
15	4	1	0

In 2018, when more than 400,000 Korean tourists visited Tsushima, a phenomenon, later called "overtourism," began to occur frequently in Tsushima. For example, Korean tourists cannot eat lunch and dinner because there are not enough local restaurants, which are always full of customers. Many Korean tourists waiting for their return ferries inadvertently cross the road in front of the ferry terminal or go off the road in a group, obstructing local traffic. Korean tourists occupy seats on public local buses, and local children and older people in Tsushima cannot get on the bus. Since Japan and South Korea have different traffic rules, accidents involving Korean tourist drivers occur frequently. Korean tourists generate a large amount of garbage. Korean financial capital has bought land and buildings on Tsushima Island, and Koreans are running hotels and guesthouses targeting Korean tourists. The fact that a small island of less than 30,000 people accepts a relatively large number of Korean tourists has created an "over capacity" challenge for the island. As a result, the residents' impression of Korean tourists has soured, and some locals have started to wonder whether Tsushima could continue to accept Korean tourists.

Then, since late 2018, political and diplomatic relations between Japan and South Korea deteriorated rapidly. The problem of Japan's responsibility for the impacts of colonial rule and discrepancies in each other's historical perceptions came into focus. Hostile measures were taken by the governments of both countries. Japan-South Korea relations worsened to the point that they were called "the worst since the end of World War II." In South Korea, the boycott of Japanese products and the movement to refrain from traveling to Japan have accelerated, and the number of Korean tourists visiting Tsushima declined sharply, approximately from 410,000 in 2018 to 260,000 in 2019. The towns of Tsushima and their residents, booming with the bustle of Korean tourists, suddenly found themselves in a dire economic situation. Still, many residents believed that diplomatic relations between Japan and South Korea would improve soon and that Korean tourists would quickly return to Tsushima.

However, an even more unfortunate event—the COVID-19 pandemic—occurred. The pandemic allowed no one to cross the border to travel to Japan. In March 2020, all international ferry services between Tsushima and Busan were suspended. It was a de facto "closure of borders." Until the reopening of the sea route between Tsushima and Busan in February 2023, the number of Korean tourists coming to Tsushima had remained almost zero for three years. As a result, Tsushima's economy and people's livelihoods were devastated. The impacts on the residents in Tsushima engaged in tourism, such as accommodation, transportation, and restaurants, were particularly severe, and many local folk either changed their occupations or had no choice but to leave the island. They have expressed strong remorse for their considerable dependence on Korean tourists before COVID-19. They have realized that if diplomatic relations between Japan and South Korea deteriorate, they cannot make a living by tourism targeting Korean tourists.

On the other hand, according to the author's fieldwork from 2020 to 2022, many island residents still admit that border tourism by Koreans is an essential part of the island's sustainability. With the reopening of international ferry service between Tsushima and Busan in February 2023, Tsushima is now gradually regaining its vitality due to the acceptance of Korean tourists. The number of Korean tourists who visited Tsushima in 2023 exceeded 120,000 (Immigration Services Agency, Government of Japan).

There are bright signs for the sustainability of Tsushima with the end of the COVID-19 pandemic and the reopening of the Tsushima-Busan ship route. Nevertheless, an event has suddenly occurred that could disrupt the lives of Tsushima islanders. Whether it is right or wrong to accept, nuclear waste disposal on the island became a hot topic in 2023. Tsushima suddenly attracted attention as a candidate site for the underground disposal of highly concentrated radioactive waste generated from nuclear power plants in Japan. The Japanese government and the Nuclear Waste Management Organization of Japan (NUMO), which is responsible for the final disposal of nuclear waste in Japan, have been looking for suitable sites for the (allegedly) safe disposal of high-grade radioactive waste. Industry groups in Tsushima, such as the "Tsushima Chamber of Commerce and Industry" and "The Cooperative Association for Construction Industry in Tsushima," acceded to the government's request to cooperate in inviting a disposal candidate site. The background is that, over the past few years, the deterioration of relations between Japan and South Korea and the COVID-19 pandemic have disrupted the number of Korean tourists and damaged Tsushima's economy. Suppose Tsushima City accepts the first phase of the preliminary investigation to determine whether it is suitable as a potential disposal site. In that case, the Japanese government will provide subsidies of up to 2 billion yen ($14 million) over two years. This amount is just about one-quarter of the approximately 9.1 billion yen ($63 million) spent by South Korean tourists in Tsushima in 2018. If the city moves to the next research stage, it can get a larger subsidy; but it might lose its appeal to both residents and visitors.

In response to requests from local industry groups that expect subsidies to revitalize the island, the Tsushima City Council narrowly decided to accept the first phase of the preliminary 'nuclear waste' investigation. However, there was also widespread public outcry against the acceptance of the investigation. Not only does the island public doubt the safety of underground disposal of highly concentrated radioactive waste, but it also fears that if their acceptance of nuclear waste is made public, then

South Koreans—who are highly sensitive to the issue of radioactive contamination—would never visit Tsushima. Under these circumstances, the mayor finally issued a statement refusing to accept the preliminary investigation, and the problem seems to have calmed down for the time being. In the March 2024 mayoral election, the current mayor, who announced his refusal to accept nuclear waste, was re-elected by a large margin (The Asahi Shimbun 2024). However, the issue has not yet been completely settled. To ensure sustainability as a remote border island, will Tsushima successfully continue to develop border tourism while overcoming the temptation to live as an "island of nuclear waste"?

5 Conclusion

A secure technology for the disposal of highly concentrated radioactive waste has not yet been established. Since the level of radioactivity of highly radioactive waste is extremely high, it is said that the only way to dispose of it is to isolate it deep underground for 100,000 years. In Japan, where earthquakes and other disasters occur frequently, an acceptable candidate site would be a "frontier" site far from metropolitan cities such as Tokyo and Osaka. In this sense, it is not without reason that Tsushima, a remote border island, is being discussed as a candidate site for receiving such toxic waste. Even in Europe, where countries share land borders with their neighbors, many nuclear power plants are deliberately located in border areas, away from population centres (Kaijser and Meyer 2018). Japan is not the only country that wants to have nuclear power plants and radioactive waste disposal sites on its "frontier."

However, the mindset of placing dangerous and nuisance facilities on the "edge" is precisely an outcome of falling into the "territorial trap" in which the territory of the state is viewed as a single, homogeneous unit. The perception and practice of viewing borders or border areas as a "frontier" of a state or a "fortress" against a neighboring country rather than as a "gateway" to the neighbor have the potential to affect the lives of border people who try to ensure sustainability by regarding the borders as a "resource."

Will Tsushima thrive as a "gateway" to South Korea in order to increase its sustainability as a remote border island by developing border tourism? Or will it accept nuclear waste as a "frontier" of Japan to secure strong financial backing from the central government in Tokyo? And is there any other choice? The dynamic movement of Tsushima, a remote border island that seeks sustainability, is bound to continue.

References

Agnew J (1994) The territorial trap: The geographical assumptions of international relations theory. Rev Int Polit Econ 1(1):53–80
Brambilla C, Laine J, Bocchi G (eds) (2015) Borderscaping: imaginations and practices of border making. Routledge, New York
Cabinet Office, Government of Japan (2022) Annual report on the aging society [Summary] FY2022. https://www8.cao.go.jp/kourei/english/annualreport/2022/pdf/2022.pdf. Accessed 16 March 2024
Cooper A, Perkins C, Rumford C (2014) The vernacularization of borders. In: Jones R, Johnson C (eds) Placing the border in everyday life. Ashgate, New York, pp 15–32
Diener A, Hagen J (2012) Borders: a very short introduction. Oxford University Press, New York
Furukawa K (2020) New developments in brderland policy in Japan. Chukyo Law Review 54(1/2):27–42 (in Japanese)
Hanamatsu Y (2016) Developing border tourism in Tsushima and Busan: borders as resources of border regions. Chiri (Geography) 61(8):44–51 (in Japanese)
Hanamatsu Y (2017) Linking Tsushima to Fukuoka and Busan. In: Iwashita A (ed) Border tourism: Developing a community through tourism. Hokkaido University Press, Sapporo, pp 35–60 (in Japanese)
Hanamatsu Y (2022) Tsushima: dynamism of border island. In: Yamazaki T (ed) Doing political geographies: Methodologies of political geography. Nakanishiya Publisher, Kyoto, pp 205–216 (in Japanese)
Hanamatsu Y (2020) Border tourism. In: The editorial committee of the encyclopedia of new geopolitics (Ed) The encyclopedia of new geopolitics, pp 558–559. Tokyo: Maruzen Publishing (in Japanese)
Hirano K (2000) International cultural relations. University of Tokyo Press, Tokyo (in Japanese)
Hosaka M, Togo K (2012) Territorial issues in Japan: northern territories, Takeshima and Senkaku Islands. Kadokawa Publishing, Tokyo (in Japanese)
Immigration Services Agency, Government of Japan (undated). Immigration control statistics Table. https://www.moj.go.jp/isa/policies/statistics/toukei_ichiran_nyukan.html. Accessed 16 March 2024
Iwashita A, Hanamatsu Y (eds) (2014) Promoting tourism on the border island of Tsushima. Hokkaido University Press, Sapporo (in Japanese)
Iwashita A (2016) Introduction of border studies: territory, sovereignty, and ideology. Tokyo: ChuoKoron-Shinsha (in Japanese)
Japan Ministry of Health (2020) Labour and welfare, population and demographics statistics (in Japanese) municipalities: 2013–2017. https://www.mhlw.go.jp/toukei/saikin/hw/jinkou/other/hoken19/dl/2-3P.pdf. Accessed 09 January 2024. (in Japanese)
Kaijser A, Meyer J (2018) Nuclear installations at the border: transnational connections and international implications. An introduction. J Hist Environ Soc 3(3):1–32. https://doi.org/10.1484/J.JHES.5.116793
Kawakubo F (2018) Border as an unfinished process: from classical geopolitics toward critical geopolitics. Geography [Chiri (In Japanese)] 63(3):46–53 (in Japanese)
Konishi S, Tamaki E, Yoshinaga J (2018) Biodemography of fertility in Japan. Springer, Singapore
Miyamoto T (1984) The forgotten Japanese. Iwanami Shoten Publishers. (in Japanese)
Mogiani M (2023) Studying borders from the border: reflections on the concept of borders as meeting points. Geopolitics 28(3):1323–1341. https://doi.org/10.1080/14650045.2022.2026329
Nagadome H (2009) Tsushima kokushi vol. 3: modern and contemporary edition: International exchange of war and peace. Tsushima: Korinsha Publishing (in Japanese)
Newman D (2006) Borders and bordering: towards an interdisciplinary dialogue. Eur J Soc Theory 9(2):171–186. https://doi.org/10.1177/1368431006063331
Rumford C (2006) Theorizing borders. Eur J Soc Theory 9(2):155–169. https://doi.org/10.1177/1368431006063330

Saeki K (ed) (2014) Tsushima in the middle ages: the history of Japan-Korea exchange depicted by people, things, and culture. Bensei Publisher, Tokyo (in Japanese)

Shimbun A (2023) Tsushima split over survey to host storage site for nuclear waste. https://www.asahi.com/ajw/articles/14953919. Accessed 16 March 2024

Shimbun A (2024) Island re-elects mayor opposed to nuclear waste disposal site. https://www.asahi.com/ajw/articles/15187080. Accessed 17 March 2024

Sohn C (2014) The border as a resource in the global urban space: a contribution to the cross-border metropolis hypothesis. Int J Urban Reg Res 38(5):1697–1711. https://doi.org/10.1111/1468-2427.12071

Spierings B, Van der Velde M (2008) Shopping, borders, and unfamiliarity: consumer mobility in Europe. Tijdschr Econ Soc Geogr 99(4):497–505. https://doi.org/10.1111/j.1467-9663.2008.00484.x

Spierings B, Van der Velde M (2013) Cross-border mobility, unfamiliarity and development policy in Europe. Eur Plan Stud 21(1):1–4. https://doi.org/10.1080/09654313.2012.716235

Takagi A (2023) New dynamics of border regions in Japan: shifting positionalities in the islands of Kyushu and Okinawa. Kokon Shoin Publishers, Tokyo (in Japanese)

Thompson P, Norris P (2021) Sustainability: what everyone needs to know. Oxford University Press, New York

Timothy D (2001) Tourism and political boundaries. Routledge, Oxford shire

Tsushima City (2020) Tsushima City long-term population vision. https://www.city.tsushima.nagasaki.jp/material/files/group/25/tyoukijinkobijyon-r2.pdf. Accessed 9 January 2024 (in Japanese)

Tsushima City (2021) Welfare plan for the elderly and the 8th long-term care insurance project (2021–2023). https://www.city.tsushima.nagasaki.jp/material/files/group/19/8kikaigohoken.pdf. Accessed 9 January 2024 (in Japanese)

Tsushima City (2022). *Tsushima City tourism promotion project.* https://www.city.tsushima.nagasaki.jp/material/files/group/8/8888888.pdf. Accessed 9 January 2024 (in Japanese)

Urry J, Larsen J (2011) The tourist gaze 3.0, 3rd edn. SAGE Publications, London

Van Houtum H (1999) Internationalisation and mental borders. Tijdschr Econ Soc Geogr 90:329–335. https://doi.org/10.1111/1467-9663.00074

Yamazaki T (2014) The territorial trap theory from a politico-geographical perspective. In: Iwashita A (ed) The disease of territory: A prescription for border nationalism. Hokkaido University Press, Sapporo, pp 7–26 (in Japanese)

Yamazaki T (2017) Memorandum on the appearance of geopolitics. Gendai Shisou (Revue de la pensée d'aujourd'hui), Seidosha 45(18):51–59 (in Japanese)

Dual Militarization of the Pacific Island: Continuing Colonial Transformation of Okinawa, Japan

Takashi Yamazaki

Abstract The United States has established numerous military bases on Pacific islands, such as Hawaii, Guam, and Okinawa, to ensure the effective forward deployment of its military forces and to maintain its global influence. The extended presence of U.S. forces on these islands has influenced the political, economic, cultural, and environmental aspects of indigenous island communities. Due to their insularity, these effects can significantly alter or militarize these societies. Alteration occurs in various forms, evident across physical, political, economic, and cultural domains. This chapter examines the interconnected dual processes of militarization in Okinawa since 1945, asserting that these processes are inextricably linked to de-militarization efforts, including local protests and both national and local initiatives aimed at reducing military bases. The de-militarization of Okinawa remains a critical focus for both the U.S. and Japan, shaped by their geopolitical aspirations in East Asia and mutual apprehensions regarding a "rising China." Currently, the Japanese government advocates for a combined approach of de-militarization and re-militarization to address ongoing local protests, while also acknowledging the necessity of the Japan-U.S. alliance. The geographical manifestation of this duality of militarization is evident in Okinawa. This chapter examines the connections between the processes of de-militarization and re-militarization in Okinawa, portraying the U.S. military's presence on these 'peripheralized' islands as an imperialist endeavor that sustains its hegemonic power while preserving its alliance with the host nation, ultimately undermining the autonomy of the islanders.

Keywords De-militarization · Re-militarization · Island · Pacific Ocean · Okinawa

---

T. Yamazaki (✉)
Graduate School of Literature and Human Sciences, Osaka Metropolitan University, Osaka, Japan
e-mail: yamataka@omu.ac.jp

T. Yamazaki and G. Baldacchino (eds.), *Relational Geographies of Islands in the Indo-Pacific*, International Perspectives in Geography 23,
https://doi.org/10.1007/978-981-96-7155-7_8

1 Introduction

The U.S. has built a large number of military bases (outposts) on Pacific islands such as Hawai'i, Guam, and Okinawa to secure the forward deployment of its military forces and maintain its global sphere of influence (Cooley 2008). The stationing of U.S. forces on those islands for many decades has had political, economic, cultural, and environmental impacts on indigenous island societies (Gonzalez 2013; Vine 2017; Davis 2015, 2020). Due to their small absolute size of land and population, circumscribed insular geographies, and their geo-social smallness relative to the physical size and footprint of the stationed forces, such impacts can entirely transform and militarize the societies. The processes of transformation are manifold, including the physical—such as land confiscation and base construction; the political—such as military rule and a split in local politics; the economic—such as economic dependence on military bases; and the cultural—such as the Americanization of local life (Yamazaki 2011).

As mentioned in the Preface of this book, Okinawa as an island militarized by U.S. forces exemplifies relational island geographies. Okinawa has been situated in shifting colonial and geopolitical contexts of East Asia since the mid-nineteenth century. In particular, the extra-island dynamic has been closely interrelated to the intra-island transformations mentioned above. Okinawa has never been an isolated or self-contained territorial unit but part of the regional and global geopolitical network of the modern inter-state system.

This chapter conceptualizes such transformative processes as "militarization" and selects Okinawa Prefecture, Japan, to investigate how the U.S. has militarized Pacific islands. Okinawa is a group of islands where complex processes of militarization have unfolded since 1945. Although militarization combines various physical and non-physical processes, it contains a duality or interlinkages between de-militarization and re-militarization. This chapter hopes to elucidate how such interlinkages could perpetuate the "colonization" of Okinawa by Japan and the U.S. This is the sacrifice of islands for the mainlands.

This chapter next illustrates the global expansion of the U.S. military and the geo-strategies of deploying and hosting U.S. bases in the Asia–Pacific region. Then, after describing the history of Okinawa's colonization and militarization, it explores how localities on Okinawa have been de-militarized and re-militarized. Finally, this chapter characterizes U.S. military presence on peripheralized islands as an imperialist project that perpetuates its hegemonic status and maintains its alliance with host countries at the expense of islanders' autonomy.

2 The Geopolitical and the Strategic

2.1 The Global Expansion of U.S. Military Bases

The overseas expansion of U.S. military bases is an example of U.S.-led global geopolitics involving localities and the locals. Many U.S. military bases have been built in other countries that share security interests with the U.S. According to the U.S. Office of the Assistant Secretary of Defense for Energy, Installations, and Environment (2023), there are 6,312 military bases and facilities in the U.S., and another 511 overseas facilities in other countries and a further 159 in U.S. non-sovereign territories. Figure 1 shows the distribution of all U.S. military bases beyond American soil in 2020. "Bases" in the Figure refer to 903 facilities or overseas "base sites" listed by name in the Pentagon's annual Base Structure Reports, each of which must have more than 10 acres or a value of more than $10 million. However, this "base" definition is not comprehensive, and it is presumed that the U.S. military uses 288 other small overseas facilities called "lily pads" and 48 U.S.-funded host nation bases (Vine 2023).

Figure 1 illustrates that U.S. military bases are concentrated in East Asia (Japan and South Korea), Western Europe (Germany and Italy), and the Middle East (Turkey, Kuwait, Iraq, and Afghanistan). In other words, U.S. military bases remain in the places where the U.S. military was deployed through the occupation after World War II, the Cold War—the Korean War, in particular, and the War on Terror. The Figure also shows quite a few bases on the Pacific and Caribbean islands. Postwar U.S. security policy has been premised on the strategic notion of "forward deployment," which deters threats from around the world to the U.S. far beyond its borders, rather than defending the motherland/mainland at its own doorstep. In line with this policy,

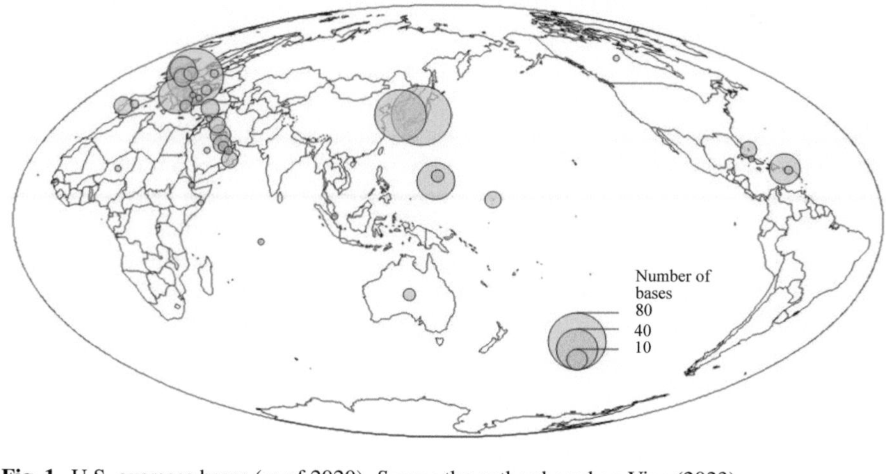

Fig. 1 U.S. overseas bases (as of 2020). *Source* the author based on Vine (2023)

the U.S. military maintains a network of forces deployed worldwide, and overseas bases are the network's strategic cornerstones.

2.2 U.S. Military Bases and Their Host Countries/Islands

The U.S. has deployed powerful forces through such a network and established and maintained its global hegemonic position. To support this network, not only the diplomatic power of the U.S. but also the agreement and cooperation of the host countries are indispensable. The attitudes of host countries are not uniform towards foreign troops' presence possessing nuclear weapons. In other words, the U.S. base policy consists of the following dimensions (Kawana 2012): the U.S. 'strategy' that takes into account threats and resource constraints; 'alliance politics' that fluctuates between the threat shared by host countries and the risk of accepting U.S. bases; and 'contracts' resulting from the negotiations and interests over bases between the two countries. Indeed, for a host country, foreign troops' presence may impair the autonomy of its security policy and increase the possibility of being attacked. On the other hand, it can also supplement the host country's military power and secure deterrence. However, as will be described below, the relationship between the U.S., the host country, and the local community that directly accepts the base is not equal. As a result, U.S. military bases worldwide have become the infrastructure that supports constructing and maintaining a world order centered on the U.S. *itself*.

A historical examination of the development of U.S. military bases in the Asia–Pacific region would help us understand how the U.S. has constructed the current world order. After acquiring California during the Mexican–American War (1846–1848), the U.S. increased its interest in Pacific affairs and repeatedly interfered with the Kingdom of Hawai'i toward the end of the nineteenth century. Winning the Spanish-American War (1898), the U.S. deprived Spain of hegemony in the Pacific Ocean and proceeded to occupy the Pacific Islands. The U.S. built military bases one after another in the territorialized Hawai'i—O'ahu Island, in particular—and later in Guam and the Philippines, which were ceded to the U.S. by the end of the 1930s. When the Northern Mariana Islands, Caroline Islands, and the Marshall Islands fell under Japan's mandate after World War I, the competition for hegemony between Japan and the U.S. in the Pacific Ocean began earnestly. In 1941, Japan attacked U.S. military bases in Hawai'i, Wake Island, Guam, and the Philippines and initiated the Asia Pacific War. In 1944, the U.S. regained Saipan and Tinian in the Northern Mariana Islands and Guam. In 1945, when the U.S. military occupied the Ryukyu and the Ogasawara Islands, it began to build large-scale bases on Okinawa Island. After the War, Japan and South Korea formed a military alliance with the U.S. They became the primary recipients of overseas U.S. military bases.

The U.S. military presence in the Asia–Pacific region is not the result of negotiations with host countries but rather the result of the expansion and confrontation of imperialist powers, including the U.S. and Japan. Here lie the origins of militarist governance. Inhabitants on the occupied and militarized islands are made to feel racially and ethnically inferior and subservient, and their human rights and political rights are not fully guaranteed. Whether or not the islands are part of a U.S. territory does not make much difference because the islands are "peripheralized" by both the U.S. and the host country (Davis 2015, 2020).

3 The Suffering: The Colonization and Militarization of Ryukyu/Okinawa

As an example of such peripheralized islands in the Asia–Pacific, this chapter selects Okinawa, the southernmost prefecture in Japan. Okinawa consists of some 300 islands, 49 of which are inhabited. Due to its location between mainland Japan and continental China, it has been influenced by both regional powers (Fig. 2). A historical overview of Okinawa helps one to better understand its geopolitical context.

In the first half of the fourteenth century, the three domains of *Hokuzan* (Northern Mountain), *Chuzan* (Central Mountain), and *Nanzan* (Southern Mountain) were formed on Okinawa Island. They entered the Chinese (Ming) dynasty's imperial tributary system at the end of the century. The system is a lord-vassal relationship with China as the suzerain. It differs from modern colonialism in that no substantial control and subordination existed in the relationship other than tribute. This system was maintained even after the three domains were unified into the Ryukyu Kingdom in 1429. In 1609, the Satsuma domain in the Tokugawa Shogunate—currently Kagoshima Prefecture, Kyushu Island, Southern Japan—dispatched troops to Ryukyu, placing

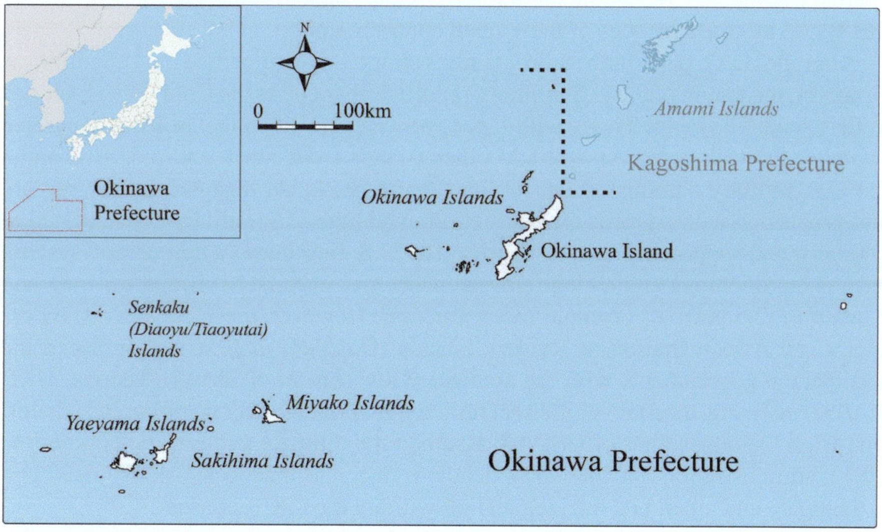

Fig. 2 Okinawa Prefecture. *Source* the author based on Wikipedia

the Kingdom under control in line with the Shogunate's intention to resume the Japan-Ming trade. The Kingdom ceded the Amami Islands to the domain and was incorporated into Japan's feudal system, while the Kingdom's regime and its relationship with China were maintained. The Satsuma domain imposed heavy tribute and trade restrictions on the Kingdom and brought it to a financial crisis. The crisis led the Kingdom to impose a poll tax on the Miyako and Yaeyama Islands, altogether called the Sakishima Islands, exacerbating the islanders' poverty.

At the end of the Shogunate period, Japan and the Ryukyu Kingdom faced Western colonialism extending to East Asia. In response, the modern Japanese (Meiji) government, newly established in 1868, incorporated the Kingdom into Japan, replaced it with the Ryukyu domain in 1872, and appointed its King as the domain load. In 1879, the Meiji government sent troops and police to Shuri, the Kingdom's capital, obliged the King to abdicate, and designated the domain as Okinawa Prefecture.[1] Japan's victory in the Sino-Japanese War (1884–1885) settled the Kingdom's status as belonging to Japan. Okinawa, however, was placed in an ambiguous position between mainland Japan and the newly colonized Taiwan. This geopolitical position brought about the following contradictions. First, the Meiji government adopted the policy to preserve Okinawan traditional customs to facilitate its control over Okinawa; but the policy delayed Okinawa's modernization. Second, the delay in modernization promoted Okinawans' domestic and international out-migration but placed the islanders as targets of discrimination by mainland Japanese. Third, to overcome the backwardness of and discrimination against Okinawans, the islanders worked to improve and assimilate their lifestyle into the Japanese cultural fabric. Finally, as Ishihara et al. (2002) demonstrate, such assimilation practices contributed to mass suicides by islanders as their self-sacrifice to the Japanese Emperor and Empire during the 1945 Battle of Okinawa. Many islanders consider the tragedies of this Battle as consequences of these deep contradictions.

After the Battle of Okinawa, the islanders temporarily welcomed the U.S. military. However, the nature of the U.S. military presence changed as the Cold War developed in East Asia. The Peace Treaty with Japan, which came into effect in 1952, separated Okinawa as well as the Amami and other Islands from Japan and placed it under the U.S. military administration. At the same time, Japan regained its sovereignty and proceeded with postwar reconstruction, while the U.S. military concentrated in Okinawa and supported Japan's security. The U.S. introduced a democratic system, such as autonomous archipelago governments and universal suffrage for men and women in Okinawa, to promote reconstruction. However, the U.S. established the U.S. Civil Administration of Ryukyu Islands (USCAR) in 1950 and replaced the archipelago governments with the areawide Government of Ryukyu Islands (GRI) in 1952, installing a system of indirect military governance over the occupied islands. The High Commissioner of USCAR assumed the ultimate authority to appoint the GRI's administration and judiciary heads, refuse and amend legislative acts, and promulgate and amend its own legislation without the GRI's consent.

[1] This forced annexation process is called *Ryukyu shobun* or the Ryukyu Disposal.

Until Okinawa reverted to Japan in 1972, USCAR lorded over Okinawan society and constrained its democracy as the U.S. prioritized Okinawa's military fortification during the Cold War. In the 1950s, the U.S. military began forcibly requisitioning private land to build new bases, but the islanders resisted in 1956 with an areawide movement.[2] The undemocratic U.S. military rule, along with the assaults and accidents by U.S. soldiers and environmental damages around the base, became factors for the islanders' systematic resistance. It developed into the large-scale reversion movement in the 1960s. On the other hand, military land rents, base employment, and reconstruction assistance measures from the U.S. military created a new class of islanders that benefited from these elements. Interests related to the U.S. military have since politically divided Okinawan society and undermined the islanders' solidarity against foreign military rule.

Okinawa was returned to Japan and became Okinawa Prefecture again in 1972. However, the U.S. military presence has continued as a result of the 1969 binational agreement. Okinawa's burden of hosting U.S. bases has significantly exceeded that of the Japanese mainland, which was drastically reduced after 1952 but ironically promoted base relocation to Okinawa. Although Japan's sovereignty and administration over Okinawa have been restored, the Japan-U.S. Security Treaty, especially the Japan-U.S. Status of Forces Agreement (SOFA), protects U.S. military bases and personnel under extraterritoriality. However, inequality in bilateral authority persists, especially over the handling of U.S. military-related incidents and accidents. After the reversion, the Japanese government promoted Okinawa's development through the vast and long-term public works investment facilitated by the Okinawa Promotion Development Plan (1972–2001). Nevertheless, the developmental investment did not necessarily lead to Okinawa's economic self-sufficiency. The prefecture economy still depends on external revenues and rentier income such as tourism, public works, and bases (Yomiuri Shimbun Online 2022; Bertram and Poirine 2018).

For these reasons, many islanders have continued to oppose the U.S. military presence on Okinawa. In the wake of the 1995 U.S. soldiers' assault on an Okinawan girl, two referenda were carried out, interrogating the merits of SOFA and questioning the relocation of Futenma Air Station to Henoko, a small fishery village next to Marine Camp Schwab (see Fig. 4). The 'base problem' continues to be a central issue of Okinawan society. In this way, the history of Ryukyu and Okinawa shows how the island society has been racked by and survived the geopolitical order—occupation and colonization—created by the major powers.

[2] The movement has been called *Shimagurumi toso,* or the island-wide struggle.

3.1 *The De-Militarized and the Re-Militarized*

3.1.1 Binational Agreements

Enloe (2000) defines militarization as a set of dynamic processes to sustain militarism. It can include not only armament but also a combination of different but inter-related—physical, economic, social, and political—processes affecting civil society (Yamazaki 2011). Militarization may manifest itself spatially to construct particular geographies and landscapes (Woodward 2004). This chapter employs another way to articulate the complexities of militarizing processes by adding two sub-categories to the concept of militarization: de-militarization and re-militarization. This conceptual addition helps us better appraise militarization's complexities as dual or interlinked processes. The processes of dual militarization may not only result from financial limits to such issues as armament, the relocation of military installations, and the innovation of military technologies but also reflect changes in the geopolitical context (i.e., the spatial distribution of military tensions and reliefs) in which the processes unfold. Although there can be new militarization somewhere in the world, the process may be related to de-militarization elsewhere as part of U.S. military bases' global deployment. In this sense, the construction and closure of bases can be inter-related at the global scale. Before analyzing the complex aspects of militarization developing on Okinawa, this section first examines how Japan was de- and re-militarized after World War II.

The basis of Japan's postwar security policy was constructed several years after the War. Japan was disarmed in 1945, began to be rearmed after 1950, and has maintained a military alliance with the U.S. since 1952. As a 'lightly armed' state with a pacifist constitution, Japan placed itself under the 'nuclear umbrella' of the U.S. during the Cold War (1947–1991). Japan established Self Defense Forces (JSDFs) in 1954 to collaborate with U.S. forward deployment against the Communist bloc in Asia. Even after the Cold War, closer collaboration between U.S. forces and JSDFs and Japan's dependence on the U.S. military presence have continued to deal with "threats" from East Asia.

The processes of de/re-militarization through the U.S. military presence have particular geographical aspects in Japan. The postwar spatial restructuring of U.S. military forces within Japan underwent the following stages. First, after the U.S. occupied Okinawa in 1945, Okinawa was separated from mainland Japan and re-militarized under the U.S. military administration from 1952 to 1972. Second, nation-wide protests were organized against the Japan-U.S. Security Treaty in the 1950s to 70s so that the area of U.S. bases was reduced in mainland Japan and partially relocated to Okinawa (Fig. 3). Third, while the Japanese public supported the Japan-U.S. alliance from the 1980s onward, the gravity of defense shifted in the 1990s from the north (against the former USSR) to the west and south (against North Korea and China). Fourth, the U.S. military's global transformation led to the incremental reduction and relocation of U.S. bases within Okinawa in the 2000s. Finally, the most recent shifts in Okinawa's militarization have appeared in the southernmost

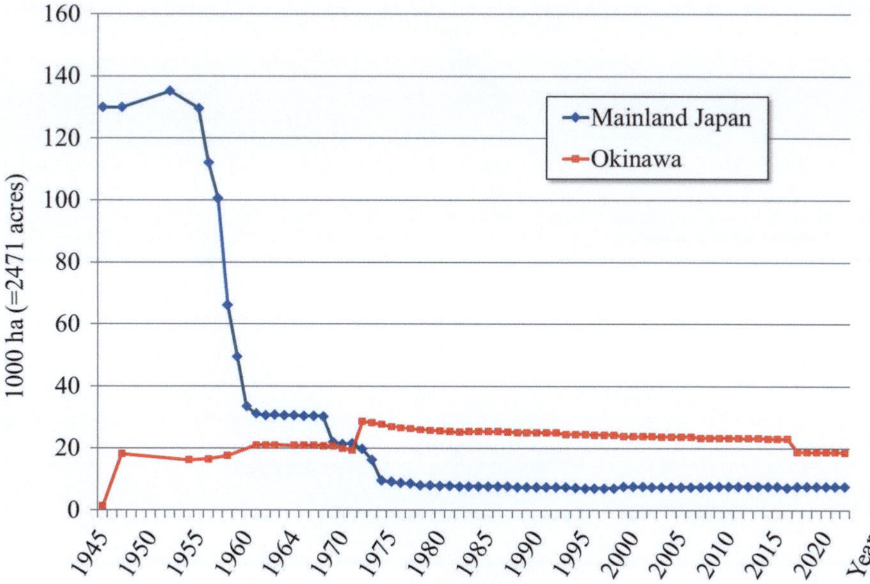

Fig. 3 The land area of U.S. military bases and installations, 1945–2022. *Sources* Okinawa Prefecture, U.S. military and Japan's SDF bases in Okinawa (each year) and other sources

group of islands in Okinawa, where new JSDF bases have been set up to counter China's movements towards the disputed Senkaku (Diaoyu/Tiaoyutai) islands and the Western Pacific (e.g., Baldacchino 2017). This current situation seems to publicly reaffirm the significance of the U.S. military presence on Okinawa and re-situate Okinawa in the context of the "New Cold War" between the U.S. and China.

Since the main drive of militarization on Okinawa has been the physical presence of U.S. military bases, tracing spatio-temporal shifts in the area of U.S. bases on Okinawa would be an objective and convenient way for analysis. The lands used or leased for bases on Okinawa increased in area until 1972 and were only incrementally returned until recently (Fig. 3). When U.S. soldiers raped a 12-year-old Okinawan girl in 1995, this incident triggered large-scale protests in Okinawa. It led to a series of tense exchanges between the governor of Okinawa and the Japanese government. The Japanese government also started negotiations with the U.S. in the Special Action Committee on Okinawa (SACO). They agreed that 5,002 ha (12,361 acres) of the land on Okinawa would be returned from 11 U.S. military installations. One of the focal points of this agreement was the closure and relocation of Futenma Air Station (481 ha). In the following years, however, only 900 ha of scattered lands (including those unrelated to the agreement) were returned. Meanwhile, a military helicopter crashed at Okinawa International University next to Futenma Air Station in 2004, making the swift relocation of the Station inevitable (Brooks 2010).

It was not until the end of 2016 that the U.S. returned 3,987 ha more land. As of the beginning of 2021, 25 years after the agreement, 4,411 ha of the initially planned

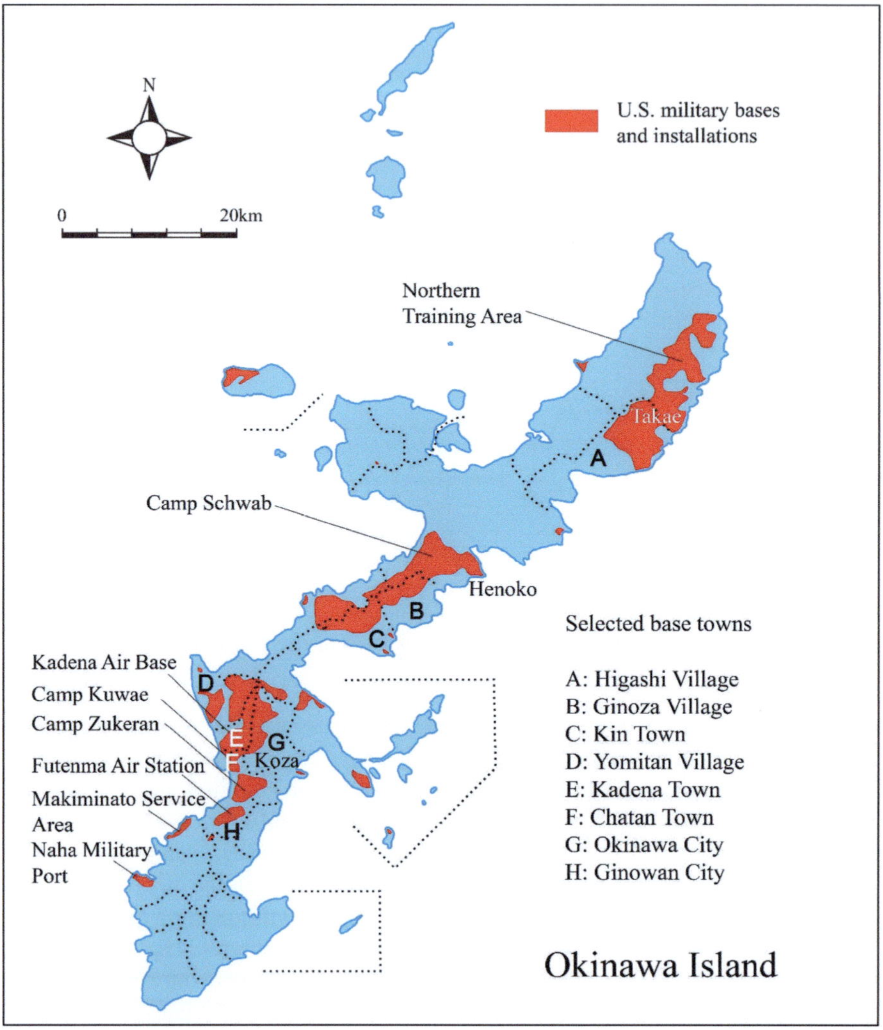

Fig. 4 U.S. bases, installations, and training sites and selected base towns (as of 2013). *Sources* https://www.pref.okinawa.jp/site/chijiko/henoko/english.html and the author based on Wikipedia

5,002 ha had been returned. Most were forests in the Northern Training Area (Ryukyu Shimpo 2021, January 5). As shown below, other base return or closure plans have been delayed because the pertinent bases have not been relocated as planned. In other words, de-militarization as base reduction or closure is to be offset by, or interlinked to, re-militarization somewhere else, mainly within Okinawa.

In 2016, ten years after the SACO agreement, the U.S. and Japanese governments approved recommendations for realigning U.S. forces in Japan and related JSDFs.

The joint statement "United States-Japan Roadmap for Realignment Implementation"[3] comprised six initiatives. Among them, "Realignment on Okinawa" proposes the following four plans:

(a) Futenma Replacement Facility (FRF)

This plan reaffirms that the FRF will be located in the Henoko-saki and adjacent water areas of Oura and Henoko Bays and include two runways aligned in a V-shape. Each runway has a length of 1,600 m plus two 100 m overruns. This relocation plan has been a target of persistent prefecture-wide protests for years, although landfill work at the construction site started in 2018. Only 12.4% of this project had been completed as of October 2022 (Okinawa Prefecture 2024).

(b) Force Reductions and Relocation to Guam

This plan proposes that approximately 8,000 U.S. Navy personnel and their 9,000 dependents would be relocated from Okinawa to Guam by 2014 and that Japan would cover part of the expenses for constructing new facilities. Although the Japanese and U.S. governments agreed to separate this plan from the FRF in 2012, the plan will begin in 2024 (Japan Ministry of Defense 2024).

(c) Land Returns and Shared Use of Facilities

Following the relocation to the FRF, the return of Futenma Air Station, and the transfer of Marine personnel to Guam, this plan recommends that the remaining facilities and areas on Okinawa be consolidated, thereby enabling the return of significant land areas south of Kadena Air Base. This plan recommends the total or partial return of the following six candidate facilities (For the returned land as of March 31, 2022, see Okinawa Prefecture 2023: 65–78 and Fig. 4):

Camp Kuwae: Total return (linked to the FRF by relocating civic facilities in the Camp). 45.0/112.4ha (40%) have been returned.

Camp Zukeran: Partial return and consolidation of remaining facilities and infrastructure to the extent possible. 237.0/771.2ha (30.7%) returned.

MCAS Futenma: Total return (linked to the FRF). 22.0/497.8ha (4.4%) returned.

Makiminato Service Area: Total return. 11.2/278.7ha (4.0%) returned.

Naha Port: Total return (relocated to the new facilities, including additional staging area, to be constructed at Urasoe). 25.1/810ha (3.1%) returned.

Army POL Depot Kuwae Tank Farm No. 1: Total return: 0/16ha (0%) returned.

This plan states, "All functions and capabilities that are resident in facilities designated for return, and that are required by forces remaining in Okinawa, will be relocated within Okinawa. These relocations will occur before the return of designated facilities." Therefore, we should note that without any feasible plan for relocation "within Okinawa," these bases and facilities would not be returned.

[3] See https://www.mofa.go.jp/region/n-america/us/security/scc/doc0605.html Accessed 18 May 2024.

(d) Relationships among Initiatives

This plan best represents the problematic nature of the realignment. It emphasizes the following three points. First, the Okinawa-related realignment initiatives are interconnected within the entire package of initiatives, and close linkages are established between them. Second, the consolidation and land returns south of Kadena Air Base depend on completing the relocation of Marine personnel and their dependents to Guam. Third, the Marine relocation then depends on the completion of the FRF and Japan's financial contributions. As mentioned above, since the construction of the FRF has been significantly delayed by continuing Okinawan protests, lawsuits, and other problems, the interlinked realignment initiatives would not be realized sooner. The 2012 agreement on this matter thus aimed at delinking the FRF from the rest of the initiatives.

In 2013, the Japanese and U.S. governments again agreed on the "Consolidation Plan for Facilities and Areas in Okinawa" to implement the "Realignment" plans (MOFA 2013). This classifies areas for the return south of Kadena Air Base into three: "Areas Eligible for Immediate Return Upon Completion of Necessary Procedures," "Areas Eligible for Return Once the Replacement Facilities in Okinawa are Provided," and "Areas Eligible for Return as USMC Forces Relocate from Okinawa to Locations Outside of Japan." The first category newly refers to four areas, which amount to only 65 ha that have been returned. Otherwise, the offsetting nature of the realignment is similar to the previous plans.

3.1.2 The Local

The U.S. military presence on Okinawa under the binational security arrangements has been so long that it has engraved various military-related incidents on Okinawans' experiences and memories. U.S. military bases occupy 14.6% of Okinawa Island's area as of 2022 (Okinawa Prefecture 2023: 5). The land for bases has been only incrementally returned to Okinawa, as seen above. Militarization has not only affected Okinawan perceptions and material lives but also influenced redevelopment plans in *kichi no machi* or base towns. Some base towns have implemented redevelopment plans for the returned land. Investigating base towns would reveal the local effects of de/re-militarization. For such investigation, this chapter has selected all eight towns in Okinawa where more than 30% of their land has been occupied by U.S. military bases and facilities as of 2010 (Table 1, Fig. 4). The amount of land occupied by U.S. bases has not been drastically reduced in these base towns since the 1972 reversion, which has been a basic political-economic condition to influence the livelihoods of town people.

A cluster analysis was conducted for these base towns using eleven political-economic variables: two for tendencies in gubernatorial election results,[4] two

[4] Traditionally gubernatorial elections in Okinawa represent a bipartisan competitive political cleavage between conservative and reformist (leftist) parties. This cleavage has been constructed predominantly over the U.S. military base issue. For most Okinawa elections, reformist votes are

Table 1 Political-economic profiles of selected base towns as of 2010 *Sources* Okinawa Prefecture annual statistics book (each year) and other sources

Base town	Variable 1	2	3	4	5	6	7	8	9	10	11	Cluster type
	Majority vote for conservative candidates (1972–2010)*	Swing** in vote (1972–2010)	Land occupied by U.S. bases (2010)	Base-related revenue rate (2010)	Population (2010)	Population density (2010)	Population increase (2005–10)	Private residential land (2010)	Employees in the construction industry (2010)	Employees in the lodging, food, and beverage service industry (2010)	Annual average of economic growth rate (2001–10)	
Unit	Frequency	Frequency	%	%		per km^2	%	%	%	%	%	
Ginowan City	5	4	32.4	3.7	91,928	4,666.4	2.41	45.95	9.40	7.48	1.37	I
Okinawa City	7	3	34.5	6.8	130,249	2,658.1	3.05	37.57	10.45	8.10	0.71	II
Higashi Village	10	2	41.5	2.1	1,794	21.9	−1.70	1.74	6.90	7.76	0.74	III
Ginoza Village	11	0	50.7	32.3	5,331	170.2	5.73	8.16	10.44	9.03	1.38	III
Kin Town	11	0	59.3	29.0	11,066	292.8	4.21	9.69	12.88	10.15	0.99	III
Kadena Town	7	5	82.5	23.3	13,827	919.3	1.45	10.32	12.55	8.40	0.58	III
Yomitan Village	1	2	35.8	12.0	38,200	1,086.2	2.40	18.66	11.00	9.75	0.99	IV

(continued)

Table 1 (continued)

Base town	Variable 1	2	3	4	5	6	7	8	9	10	11	Cluster type
	Majority vote for conservative candidates (1972–2010)*	Swing** in vote (1972–2010)	Land occupied by U.S. bases (2010)	Base-related revenue rate (2010)	Population (2010)	Population density (2010)	Population increase (2005–10)	Private residential land (2010)	Employees in the construction industry (2010)	Employees in the lodging, food, and beverage service industry (2010)	Annual average of economic growth rate (2001–10)	
Unit	Frequency	Frequency	%	%		per km^2	%	%	%	%	%	
Chatan Town	2	2	52.9	12.2	27,264	1,978.5	1.55	24.84	10.02	9.15	−0.87	IV
Prefecture average	7	3	10.2	NA	1,392,818	611.9	2.29	12.88	9.17	8.09	−0.11	

Notes * Out of eleven gubernatorial elections
** Shift in the majority vote between conservative and reformist candidates

for the impacts of U.S. bases, three for population features, and four for other socio-economic profiles (i.e., urbanization, local principal industries, and economic growth). These variables indicate the political-economic situation of the base towns in 2010, which can be understood as the foundation of their subsequent situation to the present. The results of the analysis classified eight base towns into four types (Table 1).

Types I and II represent cities in the central part of Okinawa Island, with larger populations and slight differences in financial dependence on base-related revenue[5] (Variable 4) and economic growth (Variable 11). Political attitudes towards base issues had been ambivalent, reflected in inconsistent popular support for conservative candidates in the post-reversion gubernatorial elections. Type III base towns are rather clearly differentiated from other types and show a higher occupation ratio by and financial dependence on bases (Variables 3 and 4). They are less urbanized (Variables 5, 6, and 8) and mostly located in the northern part of Okinawa Island. These geographical features overlap with higher support for conservative (pro-U.S. base) prefecture politics (Variables 1 and 2). Among Type III base towns, Kadena Town occasionally supported reformist or anti-U.S. base candidates despite its high dependence on base-related revenues. Type IV base towns are more urbanized and prone to support non-conservative candidates than Type III towns.

From these, we can say that urbanized base towns tended to support non-conservative candidates, probably because they did not depend so much on base-related revenue due to other solid income sources. On the contrary, the economic weakness of rural base towns has promoted public support for conservative politics and hosting U.S. bases. Hypothetically speaking, as Yamazaki (2008) argues, if urbanization and population growth accompany the economic growth of base towns, their motivation to maintain bases would lower. Chatan and Yomitan (Type IV) have promoted substituting economies through commercial and agricultural development measures using returned lands. In contrast, Kin (Type III) has depended highly on base-related revenue and land rents from Camp Hansen with few economic options. Unlike these smaller base towns, Ginowan City (Type I) and Okinawa City (Type II) have long dealt with base issues, such as accidents, crimes, and daily interactions, related to Futenma Air Station and Kadena Air Base, respectively, while promoting their own urban (re)development projects.

How to deal with U.S. bases has been one of the primary political agendas for Okinawa Prefecture and its municipalities. The politics of base towns has not been uniform. Instead, the bases' ambivalent (both beneficial and harmful) nature has tended to politically divide local communities over base issues, as reflected in repeated swing votes (see Variable 2 in Table 1). Such division has, in turn, hindered

---

generally cast for candidates and parties opposing the U.S. military presence on Okinawa, while conservative votes are for those accepting it.

[5] "Base-related revenue" means revenues provided by the Japanese government for compensating municipal revenue losses and preventing environmental damages caused by U.S. bases. It also includes national general grants for municipalities hostimg military bases. Other than public revenue, the Japanese government pays for local landowners the rents of U.S. bases. U.S. bases directly provide local redsidents with job opportunities and other income sources.

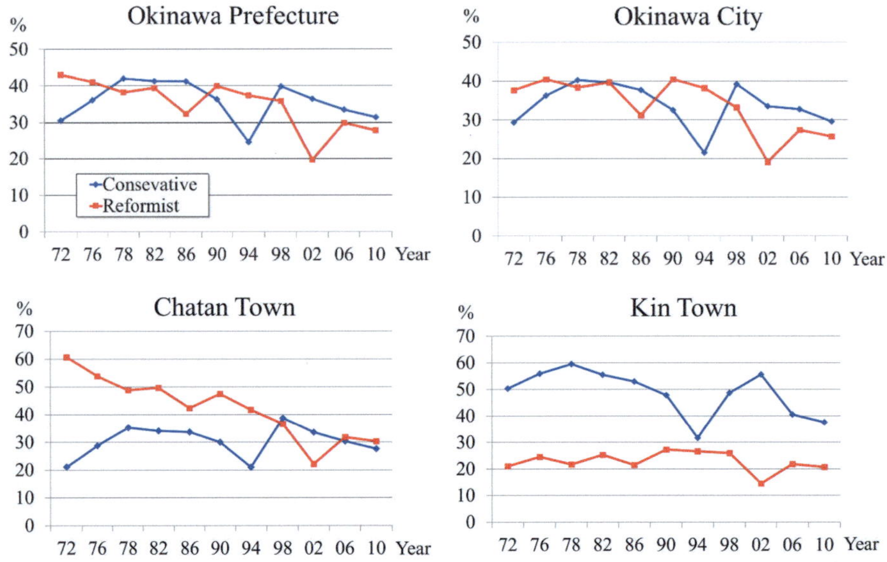

Fig. 5 The results of gubernatorial elections for selected base towns (1972–2010). *Sources* Okinawa Prefecture Election Commission https://www.pref.okinawa.lg.jp/site/senkan_i/index.html and other sources. *Note* % is the ratio of votes to the number of voting rights holders on election day

the concerted protest of the islanders against U.S. military presence and Japan's associated security policies. As shown in Table 1, urbanization, lower dependence on base-related revenue, and larger population size correlate with anti-base sentiments expressed through non-conservative votes. The gubernatorial election results (Variables 1 and 2) also show that voting patterns of these base towns can be classified into three: swing (Types I and II), conservative dominant (Type III), and reformist dominant (Type IV) votes. Figure 5 depicts these voting patterns for three sample base towns compared to the overall swing-vote pattern of Okinawa Prefecture.

The patterns expose some important underlying voting behaviors. The graphs show general downward trends, illustrating a constant decrease in voter turnout. The trends accelerated with the decline of reformist votes through the 1990s.[6] However, the weakening of anti-base politics promoted a realignment of political parties in Okinawa, forming a new political organization called "All Okinawa" for the 2014 gubernatorial election. All Okinawa has attempted to transcend the traditional conservative vis-à-vis reformist cleavage of prefecture party politics and unite oppositional forces against the relocation of Futenma Air Station to Henoko. The movement was extended to the subsequent gubernatorial and other local and national elections in Okinawa. In recent years, as Table 2 shows, while each base town still maintains its

[6] Figure 5 employs an absolute rate of votes obtained by candidates. The rate stands for the percentage of votes cast from the number of registered voters for the specific election date and reflects the voter turnout for that election.

unique political attitude towards bases, many of the towns appear to share similar sentiments against the relocation of Futenma Air Station or the binational agreements to continue militarizing Okinawa in the previous decade. However, population decline and the severe impact of COVID-19 in less urbanized (Type III) base towns show a swing to conservative politics for the status quo of U.S. militarization in the 2022 gubernatorial election.

On the other hand, an examination of redevelopment plans for base towns would help us understand the local processes of de-militarization (Yamazaki 2008). These towns do not simply accept bases as they are. Type II towns, in particular, have negotiated base reductions with U.S. forces and the Japanese government while implementing redevelopment plans for returned lands. Chatan is often seen as the first and "successful" case of commercial redevelopment on the returned and reclaimed land in which commercial and resort complexes such as "Hamby Town" and "American Village" have been built. Yomitan can be characterized by persistent anti-base sentiments among residents. The village has been promoting agricultural or endogenous development for the former U.S. military airfield site.

On the contrary, the city center of Okinawa City (Koza District) has suffered an urban decline. Its physical expansion is limited by Kadena Air Base, which attracted the local population to work for the base and its service demand in the past. Motorization and suburbanization have moved citizens away from the Koza District. Thus, the urban redevelopment plan of Okinawa City focuses more on its cultural resources (i.e., Americanized local culture and landscapes as a base town; see Yamazaki 2010).

Among the selected base towns, Kin is most financially dependent on base-related incomes and has accepted the relocation of military installations from other base towns such as Chatan and Yomitan. Although there has not yet been any growing industry substituting for base-related incomes, the town is attempting to carry out a new resort development plan for the returned land like other base towns.

Whereas the relocation of Futenma Air Station to Henoko has attracted opposition from inside and outside Okinawa, the re-militarization of the northern half of Okinawa Island has been underway to offset the reduction and closure of bases or de-militarization of the southern half. Another notable grassroots opposition has emerged against the construction of new helipads at Takae in Higashi (Type III in Table 1) to offset the area reduction of the Northern Training Area (see Fig. 4). However, the political attitude of the majority of village residents has been conservative due to the village's socio-economic marginality (see Tables 1 and 2).

We can thus suggestively conclude that there has been an ongoing spatial restructuring of the militarized island of Okinawa and its associated archipelago. While the southern half of Okinawa Island has been urbanized and has experienced population growth after the reversion, the northern half and other remote islands in Okinawa have been less developed and peripheralized. In other words, more alternative options exist for higher land revenue in the southern half, which can push bases out of the area. On the other hand, accepting bases or base relocation has become a bargaining chip with the Japanese government for the northern half and other remote islands to sustain a livelihood, with possible associated risks.

Table 2 The recent profiles of selected base towns by type (2014–2022) *Sources* same as Table 1

Base town	Cluster type (see Table 1)	Majority vote for an "All Okinawa" candidate in the gubernatorial election			Base-related revenue rate (2020)	Population (2022)	Population density (2022)	Population increase (2015–2020)	Annual average of economic growth rate (2012–2020)*
		2014	2018	2022					
Unit		Y: yes N: no			%		per km^2	%	%
Ginowan City	I	Y	N	N	7.6	100,177	5,059.4	4.03	1.90
Okinawa City	II	Y	Y	Y	9.6	142,408	2,864.2	2.49	1.90
Higashi Village	III	N	N	N	5.2	1,616	19.7	−7.09	0.70
Ginoza Village	III	Y	Y	N	28.2	5,935	189.6	4.22	2.40
Kin Town	III	Y	Y	N	26.1	10,788	285.1	−3.79	1.40
Kadena Town	III	Y	Y	Y	18.3	13,292	879.1	−1.20	2.70
Yomitan Village	IV	Y	Y	Y	6.6	41,524	1,177.0	4.31	2.70
Chatan Town	IV	Y	Y	Y	9.3	28,322	2,036.1	−0.38	2.30
Prefecture average		Y	Y	Y	NA	1,468,634	643.5	2.37	1.40

Notes * Most of the towns had negative economic growth from 2019 to 2000 due to COVID-19

The current militarization of the Sakishima (Miyako and Yaeyama) Islands through the stationing of JSDFs is a case in point. Since the 2010s, when tensions between Japan and China escalated over the Senkaku Islands, the Japan Ministry of Defense has been bolstering defense logistics in the Nansei Islands, including the Ryukyu Islands. Responding to this new geopolitical context, municipalities in the Sakishima Islands have become more willing to accept new or reinforced JSDF bases. Unlike persistent anti-base sentiments on Okinawa Island, the political conservatization of remote islands has become apparent. This shift in local politics seems to be related to the depopulation and underdevelopment of remote islands (Yamazaki 2018). Just as in the northern part of Okinawa Island, remote islands' militarization can be a way to attract new residents, boost new service demand, and generate new revenue streams for their economy. It may also improve secure control over the adjacent sea.

In contrast, following the de-militarization of the southern half of Okinawa Island, developers' and investors' desire for higher land revenues has tended to induce excessive development or investment and accelerate intra-island competition and environmental deterioration. This "post-base problem" (Yamazaki 2008) suggests that de-militarization can be conditioned, or even determined, by the local economy distorted by the former base economy. For instance, military land rent has been politically raised to make landowners sign lease contracts, increasing the area's land value beyond its actual worth on the market.

Likewise, long-term national subsidies poured into Okinawa have promoted excessive, over-populated urbanization in the southern half. For anticipated development, the islanders may accept the northern re-militarization and heavier JSDFs' presence on remote islands. Moreover, these mechanisms may generate new conflicts between relocated bases and their operations and rural advantages, such as nature conservation, improved amenities, and cheaper housing.

4 Conclusion

This chapter has sought to examine the extra-territorial expansion of U.S. military bases as an example of how global geopolitics imbricates island localities and islanders into the hegemon's world (dis)order. The U.S. military maintains a network of forces deployed worldwide, and overseas bases are the strategic cornerstones of this network. The U.S. has deployed powerful forces through such a network and established and maintained its global hegemonic position. To maintain this network, not only the diplomatic power of the U.S. but also the agreement and cooperation of the host countries are indispensable. As Kawana (2012) points out, the continuation of the U.S. military presence in Japan in general and Okinawa in particular can be considered the result of the combination of U.S. geo-strategy, the Japan-U.S. military alliance, and more local contracts to maintain the presence in particular localities. However, the U.S. military presence in the Asia–Pacific region is not necessarily the result of negotiations with host countries but rather the expansion and confrontation of imperialist powers, including the U.S. and Japan.

In 1945, when its military occupied the Ryukyu Islands, the U.S. began to build large-scale bases on Okinawa Island. After the War, Japan formed a military alliance with the U.S. and became the primary recipient of U.S. military bases in the world. The second half of this militarization process may have been secured by national negotiations, but Japan had already militarized Okinawa Island before the 1945 Battle of Okinawa. After the Battle, the U.S. troops confiscated and enlarged Japan's army bases and built new facilities. The heavy fortification of Okinawa by the U.S. started with this occupation and profoundly transformed Okinawan society, rendering it more amenable to hosting U.S. bases during the Cold War. As shown in Fig. 3, this process continued until the 1972 reversion, whereas U.S. bases in mainland Japan were significantly reduced, though more JSDF bases were built there. After the reversion, Okinawa's de-militarization was only incremental until the 1995 rape incident that drove Japan and the U.S. to the 1996 SACO agreement.

However, although the bilateral SACO and subsequent agreements have proposed Okinawa's de-militarization, it has been severely conditioned by the relocation of existing bases and their functions, as represented by the FRF case. The FRF construction at Henoko has faced persistent opposition from the non-conservative prefecture government and islanders, as well as other technical problems. After all, such offsetting relations have contributed not to the net de-militarization but to the two states' continuing "colonial" transformations of Okinawa. This offsetting process has appeared as the spatial restructuring of the militarized island; its southern half is de-militarized while its northern half is re-militarized.

As the political-economic analyses of base towns have shown, if urbanization and population growth accompany the economic growth of base towns, their motivation to maintain bases will fall. Without any available options for economic growth, base towns would depend on base-related incomes by hosting bases. The base relocation from the south to the north on Okinawa Island also reflects such socio-economic disparities on the ground. Likewise, at the scale of Okinawa Prefecture, some municipalities in the Sakishima Islands have attracted new JSDF bases not only for border security but also for socio-economic revitalization (Yamazaki 2017). Moreover, local elections in such peripheries in Okinawa have often divided local communities over base issues, making it difficult for them to collectively oppose the imposition of military burdens.

In conclusion, U.S. military bases located in the "periphery" within and out of the U.S. may remain regardless of various negative consequences on and protests from host communities. This aspect of overseas U.S. bases suggests their imperialist nature. The Okinawa case so far discussed may make us think that island militarization might be structurally perpetuated as an ongoing "colonization" process or sacrificing islands for mainlands. However, many bases have been closed so far due not only to the U.S. or host country's policy but also to the local opposition, as seen in Vieques Island in Puerto Rico and Maehyang-ri in South Korea (Hayashi 2012). There are also criticisms from a realist standpoint on the "forward deployment" strategy (Glaser 2017). Whereas Japan has hosted the U.S. military at a considerable cost (Kawase 2011), the U.S. military presence in Japan has become a strain on U.S.-China relations in recent years (Nathan and Scobell 2012; Ryukyu Shimpo 2021, December 9). If international affairs and military technology change in the future, the functions and needs of U.S. military bases may change. In other words, since bases are by no means given or inevitable but are variable, their reduction or elimination needs to be recognized as a viable policy option for host countries and communities. On such recognition, host countries could envision and shape independent security policies for themselves, thus seizing their own future.

References

Baldacchino G (ed) (2017) Solution protocols to festering island disputes: 'Win-win' solutions for the Diaoyu/Senkaku islands. Routledge, London

Bertram G, Poirine B (2018) Economics and development. In: baldacchino G (ed) The Routledge international handbook of island studies: A world of islands. Routledge, London, pp 202–246

Brooks WL (2010) The politics of the futenma base issue in Okinawa: relocation negotiations in 1995–1997. Asia-Pacific Policy Paper Series, 10. https://www.academia.edu/11837513/The_Politics_of_the_Futenma_Base_Issue_in_Okinawa?auto=download. Accessed 6 October 2024

Cooley A (2008) Base politics: democratic change and the U.S. military overseas. Cornell University Press, Ithaca

Davis S (2015) The empires' edge: militarization, resistance, and transcending hegemony in the Pacific. University of Georgia Press, Athens

Davis S (2020) Islands and oceans: reimagining sovereignty and social change. University of Georgia Press, Athens

Enloe C (2000) Maneuvers: the international politics of militarizing women's lives. University of California Press, Berkeley

Glaser J (2017) Withdrawing from overseas bases: why a forward-deployed military posture is unnecessary, outdated, and dangerous. Policy Anal 816:1–28. https://www.cato.org/sites/cato.org/files/pubs/pdf/pa_816.pdf. Accessed 18 May 2024

Gonzalez VV (2013) Securing paradise: tourism and militarism in Hawai'i and the Philippines. Duke University Press, Durham

Hayashi H (2012) The history of U.S. military bases: the formation and development of their network. Yoshikawa Kobunkan, Tokyo (In Japanese)

Ishihara M, Ohshiro M, Hosaka H, Matsunaga K (2002) Issues: the memories of the Battle of Okinawa. Shakai Hyoronsha, Tokyo (In Japanese)

Japan Ministry of Defense (2024) Relocation to Guam. https://www.mod.go.jp/j/approach/zaibeigun/saco/iten_guam/index.html. Accessed 18 May 2024. (In Japanese)

Kawana S (2012) Base politics: the origins of the post-war U.S. overseas bases expansion policy. Hakuto Shobo, Tokyo (in Japanese)

Kawase M (2011) The structure of the increase in the expense for the purpose of preserving U.S. Military Bases in Japan. Ritsumeikan Kiezaigaku 59(6):1094–1110 (in Japanese)

MOFA (2013) Consolidation plan for facilities and areas in Okinawa. Japan Ministry of Foreign Affairs. https://www.mofa.go.jp/mofaj/area/usa/hosho/pdfs/togo_20130405_en.pdf. Accessed 18 May 2024

Nathan AJ, Scobell A (2012) How China sees America: the sum of Beijing's fears. Foreign Aff 91(5):32–47. https://www.foreignaffairs.com/articles/china/2012-08-16/how-china-sees-america. Accessed 18 May 2024

Okinawa Prefecture (2023) U.S. military and Japan's SDF bases in Okinawa. Okinawa Prefecture, Naha. https://www.pref.okinawa.jp/heiwakichi/kichi/1017273/1017274/1025057/1022995.html. Accessed 18 May 2024 (in Japanese)

Okinawa Prefecture (2024) Henoko new base construction issues Q&A. https://www.pref.okinawa.jp/heiwakichi/futenma/1017409/1017413.html. Accessed 18 May 2024 (in Japanese)

Ryukyu Shimpo (2021, January 5) Twenty-five years since the SACO agreement: we request to lower the burden (in Japanese)

Ryukyu Shimpo (2021, December 9) U.S. congressional body referred to China's first use of nuclear weapons against U.S. military bases in Okinawa (in Japanese)

U.S. Office of the Assistant Secretary of Defense for Energy, Installations, and Environment (2023) FY2023 base structure report (BSR). https://www.acq.osd.mil/eie/Downloads/BSI/Base%20Structure%20Report%20FY23.xlsx. Accessed 18 May 2024

Vine D (2017) Base nation: how U.S. military bases abroad harm America and the world. Skyhorse, New York

Vine D (2023) Lists of U.S. military bases abroad 1776–2020. https://aura.american.edu/articles/online_resource/Lists_of_U_S_Military_Bases_Abroad_1776-2021/23857422. Accessed 18 May 2024

Woodward R (2004) Military geographies. Blackwell, Oxford

Yamazaki T (2008) Experience of militarism and regional redevelopment strategies: three "base towns" in Okinawa. Jinbun Kenkyu 59:71–96 (in Japanese)

Yamazaki T (2010) The historical representation of place in the military base town of Koza: the "reassessment" of U.S. military presence as a developmental resource. J Urban Cult Res 1:190–199

Yamazaki T (2011) The U.S. militarization of a 'host' civilian society: the case of post-war Okinawa, Japan. In: Kirsch S, Flint C (eds) Reconstructing conflict: Integrating war and post-war geographies. Ashgate, Surrey, pp 253–272

Yamazaki T (2017) The reinstitutionalization of island identities in Japan. In: Herb GH, Kaplan DH (eds) Scaling identities: Nationalism and territoriality. Rowman & Littlefield, Lanham, pp 185–201

Yamazaki T (2018) Understanding okinawa prefecture politics from a 'geopolitical' viewpoint. Chiri 63(3):38–45 (In Japanese)

Yomiuri Shimbun Online (2022, April 9). [Understanding basics] How is Okinawa's economy structured?. https://www.yomiuri.co.jp/politics/20220408-OYT1T50239/. Accessed 24 May 2024

Militarized Territoriality: Water Contamination and Military Operations in the Pacific

Tatsuki Kohatsu

Abstract This chapter examines episodes of water contamination and U.S. military operations across the Pacific, including Okinawa, Hawai'i, and the Philippines. Through the water contamination case, this study sheds light on militarized territoriality as a phenomenon in which maintaining smooth military operations as a strategic imperative influences determination and management of uncertainties (i.e., what is considered worth securing) across territories in the Pacific while obscuring accountability of managing them. This study shows that the phenomenon is prevalent in (post)colonial islands of the Pacific whose lands are occupied or used for military purposes. In doing so, the water contamination case and the concept of militarized territoriality offer critical insights into how the U.S. geopolitical maneuvering is contingent on the ways in which its strategic imperative can be achieved while prone to interfere with island life including how the environment and health of island residents can be secured and realized. Therefore, this study suggests that centering on islands and islanders is a critical approach for studies of geopolitics to further examine manifestations of the strategic imperative and enrich the island scholarship to deepen the understanding of workings and implications of geopolitical maneuvering.

Keywords Militarization · Water contamination · Territory · Geopolitics · The Pacific

1 Introduction

Geopolitical powers entail constant maneuvering in order to maintain their spheres of influence in practice. Historical and contemporary studies of imperial powers illuminate the need to pay close attention to how geopolitical expansion and maintenance takes place on the ground. (Benton, 2010; Lutz 2006, 2009). On this matter, critical

T. Kohatsu (✉)
Department of Geography and Environment, University of Hawai'i at Mānoa, Honolulu, USA
e-mail: kohatsut@hawaii.edu

T. Yamazaki and G. Baldacchino (eds.), *Relational Geographies of Islands in the Indo-Pacific*, International Perspectives in Geography 23,
https://doi.org/10.1007/978-981-96-7155-7_9

studies on militarization call for the need to examine the normalcy of the military presence and ask what enables it to deepen the understanding of how a geopolitical power works (Dowler 2012; Ginoza 2012; Gonzalez 2013; Enloe 2014).

Drawing from the critical approaches to geopolitics and militarization, this chapter examines a series of episodes regarding water contamination and military operations in Okinawa, Hawai'i, and the Philippines by considering responses of the U.S. military and local actors. This study suggests that the water contamination case sheds light on militarized territoriality. It is a phenomenon in which the maintenance of smooth military operations becomes manifest as a strategic imperative, influencing determination and management of uncertainties (i.e., what is considered worth securing) across territories while obscuring accountability of managing them. As I examine the water contamination case and militarized territoriality, this study foregrounds a spatial concept and geographical area that are often taken for granted in critical studies of geopolitics and militarization: territory and islands.

An examination of spatial politics is limited with regard to territory, compared to the spatial frameworks of landscape (Gillem 2007) and place (Bernazzoli and Flint 2009, 2010) to examine the normative dimension of militarization. Given that (extra) territorial boundaries are integral for states to install their military forces beyond their borders, the concept of territory warrants further critical attention. With this regard, Elden's (2007) reading of Foucault and conceptualization of territory is helpful as he argues that territory is a foundation on which calculations of the modern nation-state materialize. Rather than boundaries as given, this line of thinking helps to examine the management of what happens within and across territorial boundaries as an indicator of certain logic informing the working of the geopolitical power in question.

With the lens of territory, this chapter draws attention to islands as a critical site of investigation, for that the dominant geopolitical views hold strategic importance of islands (Vine 2009; Flint 2021), while they tend to marginalize islands as an empty space (Davis 2015). In doing so, this study considers islands as *focus,* rather than merely as *locus* (Ronström 2012). For a study of geopolitics, islands, and militarization, I suggest that this shift decenters the dominant focus on the continental/metropolitan states. Instead, it approaches centering on islands and islanders themselves as a critical mode of analysis to deepen the understanding of geopolitical power, informing the methodological aspect of this study to take into account the responses of island residents besides the state and military officials in examining the issue in question. Such an approach is particularly pertinent to studies of islands, given the tendencies of dominant geopolitical views to cater to their own interests while lacking critical reflections on their workings and implications.

Centering on islands, this chapter shows how the strategic imperative of maintaining smooth military operations is prevalent in the (post)colonial islands of the Pacific where lands are occupied or used for military purposes. In doing so, it provides critical insights into how the U.S. geopolitical maneuvering is contingent on ways in which the strategic imperative can be achieved. At the same time, islands as focus reveal that the militarized territoriality has adverse implications as it interferes with ways in which island life, including the environment and health of island residents, can be secured and realized. Overall, this study suggests further critical attention to island life and manifestations of the strategic imperative as a critical analysis. It can enrich the emergent scholarship on islands and deepen the understanding of workings and implications of geopolitical power that the dominant focus on continental and metropolitan states fail to fully understand in studies of geopolitics.

The next section discusses the calls for critically examining the normative dimension of militarization while considering spatial frameworks, particularly, through further engagement with the concept of territory. Additionally, the review section points out the need to look at islands as "focus," rather than "locus", given how dominant geopolitical and strategic views neglect islands as an empty space. Following the review, the empirical section turns to a series of episodes regarding water contamination in the Pacific. The case study scrutinizes how the responses of the military and local actors in the face of the water contamination provide insights into the working and implications of U.S. geopolitical maneuvering. In conclusion, the main points are summarized, and the significance of this study is considered.

2 Militarization, Geopolitics, and Islands

Empires and imperial powers are not steamrollers that can crush whatever comes into their way. Lutz (2006) observes that the notion of empire as monolithic and all-encompassing overlooks heterogeneity and vulnerability of imperial powers. Benton's (2010) historical study shows that the subjectivities of the proxy of sovereign powers and their interactions with the peoples they encountered affected the variegated legal geographies of Western imperial powers. In the contemporary context, studies of U.S. military bases show how the (extra) territorial expansion and maintenance of U.S. geopolitical power are entangled with local contexts where the U.S. military presence faces resistance (Lutz 2009). These studies illustrate the importance of examining events on the ground while investigating how powerful states expand and maintain their geopolitical powers.

To scrutinize the workings of geopolitical powers, Cynthia Enloe (2014) offers a useful perspective as to how militarization consists of various components that enable smooth military operations. Enloe (2014: 173) points out that:

> To analyze any base as if it were simply the sum of its budget, its equipment, its land, its chain of command, its legal basis, and its mission is to seriously underestimate all the power that is used to manage it, all the ideas that are devised to underpin it, and all the policies that are implemented to keep it running smoothly.

Her observation illuminates how militarization entails a normalization of norms, including militarism and masculinity, for the smooth operations of military bases. These embedded values not only shape policies, but also influence people's interactions in public and private spaces that (re)produce the hierarchical relationships between the military and host communities (Ginoza 2012; Gonzalez 2013). Such a critical perspective on militarization raises the need for unmasking the normalcy of military presence through examination of values and logics that enable it.

In this respect, geography and its allied disciplines have examined the workings of militarization, using spatial frameworks. For example, drawing on Enloe's observation regarding the normative dimension of militarization, Bernazzoli and Flint (2009) propose a comparative framework to investigate the relationship between production of place and militarism. In their study of two American towns, they suggest that the meanings that are attached to these places affect how non-military actors become embedded in reproducing militaristic values in their everyday lives (Bernazzoli and Flint 2010). Moreover, through a lens of landscape, Gillem (2007) illustrates that the U.S. military presence (re)produces the logics of militarism and capitalist consumerism in designing and planning its overseas bases.

In a nutshell, the concept of landscape sheds light on the logics of those in power to devise and implement their spatial imaginations in realizing what is perceived as necessary (Woodward 2014). Place, as a concept, describes how meanings that are attached to space affect the perceptions and behavior of people (Cresswell 2004). These critical approaches elucidate the utility of examining spatial politics as to how certain ways in which space is organized have implications to understand militarization and geopolitical maneuvering. In deploying geographical concepts to examine militarization, these approaches also demonstrate that militarization includes not only state actors, but also the experiences and practices of non-state actors on the ground (Dowler 2012; Woodward 2014).

However, the existing studies have limited engagement with territory in scrutinizing the normalcy of militarization and geopolitical maneuvering. Given that overseas military bases entail the installation of (extra) territorial presence of the military, territory deserves further attention. Territory is often associated with Sack's (1983) theory of territoriality as a strategy of control. In this conceptualization of territoriality, territory is a bounded space over which control can be enhanced. Pointing out the need to deepen the understanding of what territory is, Elden (2010) describes that territory is a historical and political concept. His genealogical examination of territory illuminates the development of the modern nation-state based on military and economic calculations with the advancement of technologies. Elden's (2007) approach to territory is helpful as a foundation on which these calculations materialize for the modern state to manage uncertainties. In this respect, territory is a particular way of spatially organizing things and people. Particularly, the spatial organization of things and people is informed by managing the uncertainties that could interrupt the perceived ideal order. Thus, drawing attention to territory in this manner asks what shapes the determination and management of uncertainties (i.e., what is considered worth securing and dealing with) as an indicator of certain logics informing the workings of state and geopolitical power, rather than viewing territorial boundaries as mere dividing lines.

This critical examination is pertinent to islands as islands and oceans are historically marginalized from continental consciousness in nation-state building (Steinberg 2009). Davis (2015) shows that dominant strategic and geopolitical thinking depicts islands as an empty space, rendering these geographical areas, the lives of people, and their agency geopolitically insignificant. The dominant geopolitical thinking perpetuates the unidirectional gazes of continental and metropolitan states on islands. Given this point, an emergent island scholarship calls for centering on the marginalized geographical areas in examining states and geopolitics (Mountz 2013, 2015). The renewed focus on islands facilitates unfixing the unidirectional analytical focus on metropolitan states and offer critical insights that the conventional geopolitical gazes overlook (Grydehøj et al. 2021). Centering islands also brings to the fore the question of how islands are conceptually understood in shaping the understanding of a given subject. As such, Ronström's (2012) suggests islands as focus, rather than locus, to scrutinize predominant representations of islands and diversity/particularity of these islands. As Baldacchino (2020: 4) puts it, "we allow ourselves to move beyond 'island as locus' where the island is an inconsequential place where things just happen to unfold, to 'island as focus' where one actively grapples with the constitution of islandness and how that may have had a bearing on what occurs there."

In a study of geopolitics and islands, I suggest that this shift from "islands as locus" to "islands as focus" brings into question the conventional geopolitical gazes in which islands are merely an empty space to hold control over. Instead, islands as focus consider islands and islanders themselves as active players and connect their critical insights on workings and implications of geopolitical powers. In doing so, islands as focus help our analytical curiosity to traverse the conventional territorial boundaries that confine islands, facilitating the critical examination of territory not as mere dividing lines, but as an indicator of logic that informs how a geopolitical power operates as discussed above. By foregrounding territory and islands, the current study seeks to deepen the understanding of geopolitics and scrutinizes what normally remains neglected from the dominant perspectives of continental and metropolitan states.

3 Methodology and Author's Positionality

With the lens of islands as focus, this chapter looks at water contamination and military operations in a series of episodes including Okinawa, Hawai'i, and the Philippines. As a departing point, it draws on Cresswell's (2004) proposition that moments of interruptions or "transgression" reveal insights into perceived normalcy. The perceived normalcy is concerned with a working of power since the maintenance of normalcy entails certain logic that informs calculations and actions to manage the perceived order. Considering the proposition, this study scrutinizes these moments as a point of investigation in the case of water contamination in the Pacific to examine the working and implications of power that normally remains unnoticed when it is working smoothly.

This study proceeds with the notion that taking into account responses of different actors enables the examination of the interruptions to scrutinize the business as usual. In other words, comparing what different groups say or do in response to a certain event of interruption helps to better understand what normally remains unnoticed. For this study, I consider responses of the U.S. military and local actors such as island residents, government agencies, and journalists in Okinawa, Hawai'i, and the Philippines. By consulting government documents, newspaper articles, and field observation, this study asks who says or does what as a response in facing the water contamination. In examining these responses, this study reveals what view shapes the management of uncertainties as to what is considered worth securing along with its implications.

In examining the water contamination case in the Pacific, I acknowledge my positionality as a male student researcher who was born and raised in Okinawa and studying in Hawai'i at the time when the water contamination became a public concern in these islands. Recognizing my positionality is to consider how my interactions and experiences informed my collection and interpretation of the materials as well as the scope of this study. By acknowledging my positionality, I wish to make it explicit that my primary focus on Okinawa and the groups in the study are not to take lightly the actions taken by other groups in Hawai'i and the Philippines. Rather, as Haraway (1988) points out, the recognition of one's positionality makes the social standing of researchers relational to one another so that further examination can be carried out. My limitations show a further possibility of islands as focus to situate islands in relation to one another in order to examine the working of geopolitical power.

4 Water Contamination in Okinawa

In this section, the responses of the military and local actors in the face of water contamination with Perfluoroalkyl and polyfluoroalkyl substances (PFAS) reveal how maintaining the smooth military operations became manifest as a strategic imperative, enabled by its extraterritorial status and the Japanese defense agency, that took precedence over the environment and health of the island residents in Okinawa. Okinawa is an island chain in the East China Sea between Taiwan and Japan. Once an independent political entity, the Ryukyu Kingdom held political authority over the archipelago. The current name, Okinawa, was given to the archipelago when the Japanese state annexed the kingdom in the late nineteenth century with an ensuing assimilation policy, including the prohibition of speaking and teaching the Ryukyuan language(s). During World War II, Okinawa became one of the deadliest battle grounds where more than 120,000 civilians—approximately a quarter of the island population, including those who were conscripted—lost their lives, and the militaries on both sides suffered devastating casualties (Ministry of Internal Affairs and Communications 2009). While Japan regained its independent status after the San Francisco Peace Treaty in 1951, Okinawa remained under the control of the U.S. military where the U.S. established a military government to make the archipelago a military outpost. In 1972, the reversion of Okinawa terminated direct U.S. military control; however, the U.S. military bases remained in Okinawa.

Since World War II, the U.S. military has maintained its bases. Approximately 70.4% of the area exclusively used for U.S. military facilities in Japan is currently concentrated in Okinawa Prefecture while the prefecture comprises only about 0.6% of Japan's total land area (Okinawa Prefectural Government 2018: 6). The U.S. military exclusively uses 15 percent of land on Okinawa Island, one of the islands in the archipelago. The high concentration of military bases has caused social and environmental problems. These problems range from violence against island residents—particularly women and children (Akibayashi 2020)—to the environmental problems that affect people's relationships to one another and to their land, including the construction of a new U.S. military facility in the northern part of Okinawa and water contamination discussed below (Oshiro 2023).

PFAS are a group of compounds that do not dissolve in natural environments; hence, they are known as 'forever chemicals.' Accumulation of the chemicals in a body raises concerns of increasing chances of health issues including cancers. Babies in their mothers' bodies are also vulnerable to the chemicals, potentially causing problems such as low weight at birth. PFAS were first detected in one of the wells inside the Kadena Air Base in 2008. After the detection, the Okinawa Prefectural Government stopped drawing water from the well while it did not pursue further investigations due to the lack of established methods for data collection (Okinawa Times 2016). More reports of high levels of the chemicals brought this matter to the public's attention in 2016. Among them, one of the primary water supply facilities in Chatan Town was particularly concerned because the facility that served roughly 450,000 residents drew water from the rivers running through and around Kadena Air Base as well as the wells that tapped into the aquifer where high levels of PFAS were detected (Mitchell 2020). At the time of the writing (September 2024), particular compounds of PFAS, called PFOS and PFOA, are subject to regulations for their production and import in Japan. For PFOS and PFOA, the current measurement is an advisory standard of 50 ng/L to show what the advised combined amount of these chemical compounds is for drinking water (Ministry of Environment 2024). Another compound, PFHxS, is a new addition to the list of PFAS compounds for regulations (Ichino 2023).

The water contamination issue brought up the Status of Forces Agreement (SOFA). The agreement came into force with the Mutual Security Treaty between Japan and the United States in 1960. The SOFA signifies extraterritorial status of the United States. The agreement gives special legal standing to the U.S. military in Japanese territory, putting the U.S. military outside Japanese jurisdiction. It has been contested for decades because the special legal rights enjoyed by the U.S. military make it difficult to deal with "spillovers" of the overseas military bases whose impacts go over the fences; these "spillovers" include noise, training accidents, pollution, and crime (Gillem 2007). The United States and Japan updated the SOFA in 2015 regarding potential environmental problems and base operations. While allowing national and local governing bodies in Japan to make a request for an on-site observation to the Commander, United States Forces, Japan (COMUSJAPAN), the agreement maintains that:

> In formulating a response, COMUSJAPAN or designee will take into account whether granting the request would interrupt military operations, compromise force protection, or hinder the management of the facilities and areas, and respond as promptly as practicable. If approved, any such observations may only be conducted in a manner that does not interfere with the actions of the United States armed forces in addressing the spill or with other operations (The United States-Japan Joint Committee 2015).

This agreement maintains its previous stance that the investigation is allowed only to the extent that it does not interfere with the base operations. In other words, the SOFA remains ambiguous in defining the responsibility of the U.S. military in dealing with potential spillovers.

With this extraterritorial status of the U.S. military and the SOFA, the responses of the military affects what can and what cannot be known regarding the water contamination. Since 2019, the Okinawa Prefectural Government and municipal governments have made requests for on-site investigations. However, the U.S. military has largely been unresponsive, with exceptions limited to incidents such as the notable leaks of fire extinguishers off base where the foam flew in the air and filled water paths in residential areas in 2020. The Okinawa Defense Bureau (ODB) under the Japanese Ministry of Defense maintains the U.S. military standing through its intermediary role. For instance, when a report of high levels of PFAS prompted the municipal government of Kin Town to request an investigation on Camp Hansen in the northeast part of Okinawa Island, the spokesperson for the OBD responded that "There is no clear evidence that the contamination comes from the base", hence no reason to put the request forward to the U.S. military (Burke 2023). As a result, the U.S. military legitimately remains unresponsive while keeping its operations running smoothly and obscuring the extent to which the uncertainties regarding the environment and health of the residents can be addressed.

In response, local residents, government agencies, and journalists make visible the chemicals that are invisible to human eyes and shed light on the relationship between the military operations and the contamination. The agencies of the Okinawa Prefectural Government and local municipal governments have conducted environmental surveys around the major U.S. military installations to detect where water sources were contaminated with PFAS since 2016. These surveys indicate high levels of PFAS around the bases including the Kadena Air Force, Futenma Air Force Bases, and Camp Hansen (Okinawa Prefectural Government 2024a; see Fig. 1).[1]

Furthermore, local resident groups facilitated off-base investigations and disseminated information by building connections with the local government agencies and media (Mori 2022). In doing so, they took up a wide range of activities to make visible the PFAS around the bases and the relationship between the water contamination and the base operations. For instance, since 2019, 有機フッ素化合物 (PFAS) 汚染から市民の生命を守る連絡会 (*yūki-fusso kagōbutsu osen kara shimin no inochi wo mamoru renraku kai*: Liaison Committee to Protect Citizens' Lives from PFAS Contamination) has held forums to share information and raise awareness of the

---

[1] Figure 1 shows how high levels of PFAS have been detected around the major U.S. bases—Kadena Air Base and Marine Corps Air Station Futenma. For more information, the PFAS data is available at the Okinawa Prefectural Government website listed in the References (in Japanese).

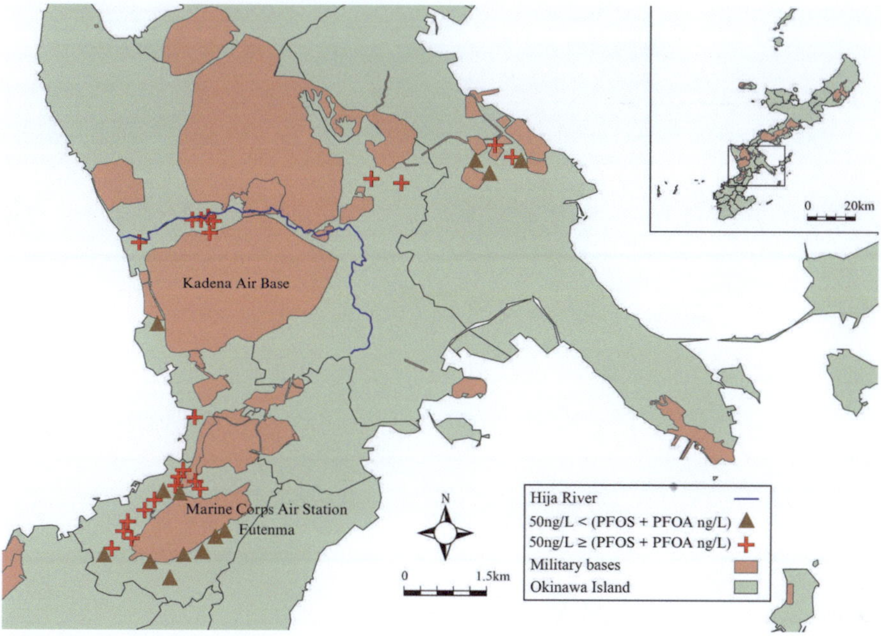

Fig. 1 Map of PFAS by the Author (Based on the PFAS Data, Okinawa Prefectural Government 2024a)

water contamination. The island residents carried out blood testing to examine accumulation of PFAS in bodies, showing higher levels than the national averages (Higa 2022). The samples from the areas with major U.S. military bases, such as Chatan Town and Kin Town, revealed numbers that exceeded a standard indicating potential health risks (Sesoko 2023). Compared to the area where there was no military base nearby, the areas with the major military bases showed high levels of PFOS detected in people's blood. In addition, a local group, called 美ら水会 (*chura-mizu kai:* Association for Clear Water: Translated by the Author), initiated soil testing to examine whether the soils near the bases had been contaminated (Arakaki 2022). Later, the prefectural government followed suit. The soil testing indicated a high level of PFAS in a school ground next to Marine Corps Air Station Futenma (Okinawa Prefectural Government 2024b).

Moreover, journalists conducted investigations to make the invisible visible. Jon Mitchell, for instance, used the Freedom of Information Act to obtain documents regarding the operations of military bases. By tracing the historical and contemporary military uses of the chemicals, Mitchell's (2020) investigation identifies three main activities on the bases in Okinawa, including firefighting training, accidental leaks,

Fig. 2 Photo of an incident on December 9, 2009 where PFAS contaminated-firefighting foam was accidentally released at MCAS Futenma, obtained through FOIA: Credit Jon Mitchell

and disposal of the chemicals (see Fig. 2).[2] His documentation of these activities provides insights into how prevalent the uses of PFAS are in U.S. military operations, further raising the need for an on-site investigation.

In short, the local actors highlight the need to deal with the uncertainties regarding their environment and health in the face of water contamination. Their responses reveal a great extent of evidence regarding the existence of PFAS in the water paths, soil, human bodies, and military operations. In relation to the local actors, the unresponsiveness of the U.S. military points to how the maintenance of smooth military operations manifests as a strategic imperative assisted by its extraterritorial status and the intermediary role of the Okinawa Defense Bureau. Thus, comparing the responses of the military and local actors illuminates how the strategic imperative of maintaining smooth military operations influences the determination and management of uncertainties as to what is worth securing while restraining the ways in which the environment and health of island residents can be secured on the island.

[2] The photo of MCAS Futenma (Fig. 2) was obtained and provided by Jon Mitchell, showing an example of the FOIA documents that reveal the military operations and its use of PFAS-contained products.

5 Water Contamination in Hawai'i

This section considers how the military responded to water contamination with jet fuel and the calls to shut down the facility associated with it in Hawai'i. The responses of the military show that the importance of maintaining its smooth operations affected how the water contamination could be handled. The parallel between Okinawa and Hawai'i further brings to light the strategic imperative of maintaining smooth military operations as well as how the U.S. geopolitical maneuvering is contingent on the ways the strategic imperative can be secured and realized.

Similar to Okinawa, Hawai'i has faced geopolitical powers, first the United Kingdom and later the United States, since the eighteenth century. As an independent political entity, the Hawaiian Kingdom established formal relations with other countries through treaties while dealing with the colonial maneuvering of the Western settlers to change the social and political relations of Hawai'i (Osorio 2002). The American capitalist and strategic interests in Hawai'i were exemplified by the Treaty of Reciprocity that fueled the plantation economy and allowed the U.S. military to seize Ke Awa Lau O Pu'uloa, known as Pearl Harbor (Kajihiro 2013: 134–135). As Kajihiro (2013) points out, the Spanish American War further laid a ground to forge the illegal overthrow and annexation of Hawai'i, enabling the American expansion to reach other islands including the Philippines and Guåhan.

Since the overthrow, the United States has occupied a large portion of land for military bases and training operations, taking up 24.6% of O'ahu along with continued expansion on other islands (Kajihiro 2013). One installation with vital importance to the military operations was an underground storage facility the U.S. military built in the 1940s at Kapūkakī, called Red Hill in English, on the Island of O'ahu. Building the underground facility was a strategic decision to maintain the military operations due to the concern about the vulnerability of the existing above-ground storage facilities at Pearl Harbor from a potential attack during World War II (Navy Region Hawaii n.d.). In this way, fuel could be secured and supplied through pipelines to continue operations. However, what was not given sufficient consideration in the calculations was the fact that underneath the fuel storage facility, lies an aquifer that is an important water source, currently serving over 90,000 residents on the island.

In 2014, approximately 27,000 gallons of fuel leaked from the storage facility (Environmental Protection Agency 2024). After the leak, the Navy and the Defense Logistics Agency made an agreement, called Agreement on Consent, with the U.S. Environmental Protection Agency and the Department of Health for continuation of the fuel storage in 2015. The procedure was to improve the facility while conducting further study to assess the fuel leak and the conditions of the facility. In 2021, there was a contested hearing that included local agencies such as Sierra Club and the Board of Water Supply, when a five-year permit was up for renewal to continue the operations at the Red Hill facility. Later, it was revealed that the Navy kept another leak a secret at Pearl Harbor during this time. In an email, a Navy captain stated that "There are significant political concerns if this were to become an 'active' leak … Activist organizations will use this to advance their anti-Red Hill narrative…at a sensitive time as the contested case hearing begins and (the) legislative season starts" (Jedra 2021). This incident was confirmed to have a significant likability of a link with the Red Hill Storage. Regardless, the need to continue the military operations prompted the internal decision to withhold the information as a Naval official communicated in an email that "This release into the harbor is not only an environmental concern but also a concern as it relates to the Red Hill fuel system" (Jedra 2021).

However, the environmental and health concerns became undeniable in the following series of reported fuels in water sources. This time, the fuel appeared in tap water and caused reported symptoms, particularly in military residential areas that drew water from the military water system (Bodon 2022). The contamination led to a temporary evacuation of residents in the military housing. In the midst of that, the Navy continued to contest the state order to remove fuel from the underground storage facility (Kim 2021). In response, groups of Kānaka Maoli and island residents in Hawai'i, such as O'ahu Water Protectors and Sierra Club of Hawai'i, pushed the military to defuel the tanks and shut down the facility. When accepting the state order, with regard to the future possibility the Navy maintained that "The decision on whether to contest the order has to take into account the strategic importance of Red Hill, the alternatives, and a number of factors that are associated with the combatant commanders' responsibilities" (McCullough 2021). The remark exemplifies how the need for the maintenance of smooth military operations persisted in dealing with the situation.

Similar to Okinawa, this water contamination case shows that the strategic imperative influenced how the situation could be dealt with. It illuminates that the military as a state apparatus manifests not only as a forward/defense force for the state(s) to deal with the potential disorder of the perceived outside. But also, it necessitates rationalization to maintain the disorder under control within the perceived inside of the territory where the military operations are concerned. Thus, as the responses of the military show, the geopolitical maneuvering entails securing the conditions in which the strategic imperative can be achieved. In doing so, this rationalization is prone to interfere with the ways in which uncertainties regarding the environment and health of people can be secured.

6 Militarized Territoriality of the Pacific

Drawing on Cynthia Enloe's definition of militarism, Teaiwa (2011) describes that militarism "'bleeds' across formal boundaries: militarism is a force that is not contained by military institutions, but one which seeps into much more fundamental aspects of social and cultural life" (p. 2). In this sense, militarism is a pervasive brief in military forces. For it to operate, it influences what is considered important of the place and people where it is concerned. Territoriality—the conditions of territory as Elden (2010) puts it—is conditional development of a territorial-reality informed by certain logic influencing calculations for management of uncertainties.

Thus, I frame militarized territoriality as a phenomenon of a territorial-reality of a given place where the strategic imperative of maintaining smooth military operations influences determination and management of uncertainties, as in what is considered important to secure.

What militarized territoriality shows, is the blurring of the conventional boundaries of the military as a state apparatus to combat the perceived enemies and maintain the order outside of its own territory. The militarization of territoriality points to the rationalization of its strategic imperative in the places where the military uses land for its operations. The strategic rationale of maintaining smooth operations permeates conventional boundaries of states as it entails keeping potential disorder (e.g., political, environmental, and infrastructural) under control for the geopolitical power to maintain itself. As shown in the previous sections, the strategic imperative particularly become apparent in the islands whose land uses are tied to the historical and contemporary colonial and militaristic interests of their metropolitan and continental states. These cases of Hawaiʻi and Okinawa reveal that the manifestations of the strategic imperative are prone to constrain ways in which the environment and health of people on these islands can be secured. Below, this section considers how militarized territoriality of the Pacific also involves postcolonial islands, such as the Philippines.

In the process of closing the underground storage facility on Oʻahu, relocating the fuel became a critical task for the military to maintain its operations by moving the fuel within the island and out to places across the Pacific including the Philippines. The United States occupied the Philippines upon the Spanish-American War in 1898 until World War II except for the years of Japanese colonization. After WWII, the United States and the Philippines made a treaty for collective defense. The Visiting Forces Agreement between the Philippines and the United States enhanced the military relations of these countries, allowing the U.S. military for its temporary visits in the Philippines while ensuring the special legal standing of the U.S. military and its soldiers on duty. Additionally, the Enhanced Cooperation Defense Agreement (ECDA) grants the U.S. military access to selected locations in the Philippines for joint training and the pre-positioning of equipment and materials in preparation for the time of military conflicts. This means that more than accommodating the physical presence of the U.S. military, the Philippines remains a part of the military logistical network for which the U.S. Defense Logistics Agency (DLA) takes on a primary role to maintain the military operations across the Pacific.

When the shipment of 39 million gallons of fuel went to Subic Bay from Oʻahu, it sparked attention. A Philippine senator raised a point that Subic Bay, a former U.S. Naval base, was not a site that "the U.S. military can use to pre-position forces or equipment under the 2023 agreement" (Knodell 2024). Subic Bay took its current form as a freeport and special economic zone under the Subic Bay Metropolitan Authority after the extension of the U.S. military use of the site was rejected in 1991. The remark referred to the revived ECDA that increased the number of sites for U.S. military use. In response, when asked about the shipment of fuel, the Philippine military stated that "the process that was followed by the U.S. government, these are all administrative in nature and did not involve the participation of the armed forces"

(Knodell 2024). While not specifying the future use of the fuel, the spokesperson at the U.S. embassy also stated that "All arrangements for the transfer and storage of this fuel were made through the proper channels, using established logistics contracts with Philippine commercial entities" (Aglibot 2024).

The contract-based logistical network has become a part of the U.S. military maneuvering to maintain its operations since the early 2010s, focusing on places such as Guam, the Philippines, and Japan. The U.S. military has refueled at Subic Bay in the past. In this scheme, the DLA facilitates the use of oil storage and logistical facilities through commercial channels, rather than merely relying on government infrastructure (Goulait 2015). Mark McLeod, the DLA Energy Commander Air Force General, is cited to state that "What's good about this economically is that we can create contracts where we need them and not have government infrastructure. …They tend to be very flexible and very affordable. If we need it, we can use it" (Goulait 2015). It enhances the flexibility of military operations to move fuel and supply it where it is needed. At the same time, this use of non-military channels for enhancement of military logistics further integrates and normalizes the military use of sites such as Subic Bay while obscuring the military relations as the responses of the state officials exemplify above. In other words, the strategic calculations of military operations become further permeated into the postcolonial relations of the Philippines with the United States for the U.S. geopolitical maneuvering to maintain its operations.

When the U.S. military was carrying out the relocation of fuel, groups in Hawai'i and the Philippines connected to raise the issue surrounding the shipment via an online forum. Women's Voices Women Speak, the group in Hawai'i, shared how the leaks of fuel had affected the island of O'ahu while pointing out the lack of transparency and responsibility on the part of the military. In doing so, the convening rendered visible the connection of the Red Hill water contamination with the shipment of fuel, rather than the shipment being merely a commercial transfer that the state officials maintained. On this matter, Stop the War Coalition Philippines contended that:

> Aside from health and environmental repercussions, the security risk is posed by who will ultimately use it, what they will use it for, and who will be responsible for securing it. …We are talking about massive amounts of dangerous, combustible substances. We demand answers, accountability and transparency in this matter that will impact the public (Aglibot 2024).

The remark brings about uncertainties regarding not only the environment and health of people, but also the question of sovereignty in terms of how these uncertainties can be dealt with and who will be responsible. The response of the group further raises the need to scrutinize how the strategic imperative permeates the territorial boundaries while creating conditions in which such objectives can be achieved. The permeation of the strategic imperative also obscures the accountability of determining and managing uncertainties, which has significant implications for the lives on these islands.

7 Island Relations Beyond Militarized Territoriality

As discussed in the previous section, centering on islands provides critical insights to deepen the working and implications of geopolitical maneuvering. In addition, islands and islanders bring to light responses to rethink the conditions of militarized territoriality. For instance, after the 2021 report on the fuel leaks, Shimanchu Wai Protectors—those who were from Okinawa and descendants of Okinawan immigrants in Hawai'i—started working to support the significance of Wai (water) for Kānaka Maoli and the lives on O'ahu. They joined rallies, helped with events, and participated in water distributions for families who had been affected by the water contamination. Especially when PFAS contamination was reported near Red Hill, Shimanchu Wai Protectors organized a rally in front of the Department of Health. Kimie Toma Shiira, a member of Shimanchu Wai Protectors, said in an interview that:

> The Okinawan diaspora stand in solidarity with our ancestral homeland as they continue to fight for the basic security of clean water and a clean environment. We need to make the U.S. military take responsibility for contaminating Okinawa's *nuchi nu miji* [the water of life]. In Hawai'i, too, there is a saying 'Ola i ka wai'—which means 'Water is life.' (Mitchell 2023)

The responses of the group brought to the fore how the military operations caused similar problems in these islands, rather than viewing the water contamination as separate instances in different places. In doing so, as Shiira's remark illustrates, the group connected Hawai'i and Okinawa by foregrounding the significance of water in relation to the prominent efforts by Kānaka Maoli and island residents for water in Hawai'i.

The importance of water engendered a critical dialogue and (re)consideration of the dominant military presence and island life. One example is an event organized by Hawai'i Okinawa Alliance (HOA). HOA consisted of Okinawan activists and scholars mostly based in Hawai'i. The event was a screening of a documentary made by Jon Mitchell, an aforementioned journalist, and Natsuko Shimabukuro, a journalist and film director for Ryukyu Asahi Broadcasting Corporation, followed by a talk-story dialogue with panelists from Hawai'i and Okinawa. Among the panelists was Chihiro Komine, an Okinawan scholar and activist. On the water contamination in Okinawa and Hawai'i, Komine pointed out how the water issue made her think about not only the military domination and control embodied by the fences, but also the disconnections that these fences had caused. As an example, she drew attention to a river, called *Hijagawa* (Hija River), that runs through the U.S. military installations in the central part of Okinawa Island. It was one of the water paths where the high levels of PFAS were detected, leading to the decision at the Chatan Water Treatment Plant to temporarily stop using the water source for drinking water (Asato 2022). As the water path ran down to the ocean through the military installations (see Fig. 1), foregrounding the significance of water brought into focus the flows of the river, the water contamination, and the land that the U.S. military presence occupied.

In thinking about the river, Komine invited Norman Kaneshiro, a third-generation Okinawan cultural practitioner and co-founder of a Ryukyu performing arts non-profit organization in Hawai'i, to introduce a song called *Hijagawa Kouta.* The song was composed by Shigeo Kyan portraying the river and island life. Though it is not clear when exactly the song was written, it is likely that Kyan composed the song after WWII. Kaneshiro sang a verse from the song (See the video: https://www.youtube.com/watch?v=CplXTmZPjYg [38:40–40:40]). The verse went as follows:

> 比謝川の水や、御万人の命、(*Hijagawa nu mijiya, Umanchu nu inuchi,*)
>
> 水の糸ちなぢ、朝も夕さも、(*Miji nu itu chinaji, Asan yusan,*)
>
> 朝も夕さも、(*Asan yusan*)
>
> Water of Hija River, Lives of Everyone,
> Thread of water weaves, Day and night,
> Day and night
> (English translation by the author)

After singing the song, Kaneshiro explained that:

> The water of Hija River is tied to the lives of everyone. This water connects us like a thread. It draws connections to us. It ties us all together and binds us together. This is perpetual, day and night. Our connections to life are our connections to each other in this way.

With the significance of water and land, Komine and Kaneshiro introduced the song from an indigenous perspective to reconsider what was rendered invisible by the military presence on Okinawa Island (Komine, Personal conversation, February 29, 2024). Komine picked this song as a means of reclaiming spaces that had been taken over and reshaped through violence and domination of the military. Violence derived from taking land as it shaped how the physical and social relations of the land to people could be understood. From this point of view, *Hijagawa Kouta* signified a memory where people's experiences and relationships to the land could be carried from the past to the present (Kaneshiro, Personal conversation, August 27, 2023). It is a non-material and non-physical space that could be accessed through mediums such as songs to consider different ways in which space could be remembered and reconsidered. Such a remembering is essential in places where physical access to the land is made impossible. Hijagawa Kouta captured the imagery and essence of this life-giving river that could not be seen today because of the physical state it is in (Komine, Personal conversation, August 27, 2023). As Kaneshiro and Komine performed and interpreted the song, their rendering of *Hijagawa Kouta* brought to light a memory about the ways that life-giving water flowed and island life connected one another, juxtaposed and contrasted with the trajectories of military control and environmental degradation charted by the water contamination on the island.

The need for remembering is telling of the conditions of militarized territoriality that restrain island life including the use of land as well as the management of uncertainties such as the environment and health of people where military operations are concerned. Through the water contamination, the dialogue emerged as a critical response, raising the need to consider that geopolitical maneuvering and military presence cause and normalize the violence of constraining how island life can be realized. As such, an act of remembering in the case of water contamination itself is a critical practice that would be otherwise ignored with the mere focus on the dominant metropolitan and continental perspective, which reinforces the normalization of the militarized conditions in return. For that, alongside other episodes above, this episode further illustrates the need and possibility of attending to islands and islanders providing critical insights regarding the working and implications of the U.S. geopolitical maneuvering.

8 Conclusion

This chapter has looked at a series of episodes regarding water contamination and the U.S. military operations across the Pacific. Centering on Okinawa, Hawai'i, and the Philippines in the case of water contamination, this study sheds light on militarized territoriality where the strategic imperative of maintaining smooth military operations influences the determination and management of uncertainties across territories while obscuring the accountability of managing them. In doing so, this study shows how manifestations of the strategic imperative are prevalent in (post)colonial islands of the Pacific where lands are occupied or used for military purposes.

Thus, the study suggests that centering on islands and islanders is a critical approach for studies of geopolitics. As this water contamination case demonstrates, it is critical as islanders and islands including lands and waters offer critical insights regarding how the strategic imperative becomes manifest. Closely examining the manifestations of the strategic imperative deepens the understanding of workings of geopolitical maneuvering. Furthermore, it is critical because failing to do so not only limits the understanding of how a geopolitical power works, but also obscures and normalizes its potential adverse implications.

This critical attention to islands is significant in the contemporary Pacific where the U.S. is seeking to enhance its geopolitical power with its allies in the competition with China. In the geopolitical situation, for instance, the U.S. military is currently enhancing its mobile and flexible operations by extending its military installations across the Pacific, including Palau and Federated States of Micronesia (Kuper 2024). At the same time, China is also turning to islands in areas such as the South China Sea (Rodenbiker 2023). The geopolitical currents further embed islands in the geopolitical maneuvering of continental and metropolitan states. For that, it is important to attend to how the strategic imperatives of these states are intertwined with islands and lives of islanders for the critical examination of geopolitics.

References

Aglibot JR (2024) Tanker carrying US Navy fuel for PH storage cancels entry to Subic. *INQUIRER.net*. https://globalnation.inquirer.net/225214/tanker-carrying-us-navy-fuel-for-ph-storage-cancels-entry-to-subic. Accessed 1 October 2024

Akibayashi K (2020) Okinawa women act against military violence: an Island feminism reclaiming dignity. Okinawan J Isl Stud 1(1):37–54

Arakaki W (2022) 米軍由来PFAS調査で土壌サンプル採取 基地に隣接する普天間第二小の周辺3カ所 ちゅら水会 [PFAS soil sampling at three sites around the Futenma Second Elementary School adjacent to the base]. *Ryukyu Shimpo*. https://ryukyushimpo.jp/news/entry-1567391.html. Accessed 29 September 2024

Asato S (2022) 北谷浄水場、PFAS低減策強化で比謝川取水の停止を継続 長田・天願川は抑制 [Chatan Water Treatment Plant continues to suspend pulling water from Hija River due to the enhanced measures to reduce PFAS. Moderation of water intake from Nagata and Tengan Rivers. *Ryukyu Shimpo*. https://ryukyushimpo.jp/news/entry-1608090.html. Accessed 1 October 2024

Baldacchino G (2020) Okinawa, Japan and our world of Islands. Okinawan J Isl Stud 1(1):1–9

Benton LA (2010) A search for sovereignty: Law and geography in European Empires, 1400–1900. Cambridge University Press, Cambridge

Bernazzoli RM, Flint C (2009) Power, place, and militarism: toward a comparative geographic analysis of militarization. Geogr Compass 3(1):393–411. https://doi.org/10.1111/j.1749-8198.2008.00194.x

Bernazzoli RM, Flint C (2010) Embodying the garrison state? Everyday geographies of militarization in American society. Polit Geogr 29(3):157–166. https://doi.org/10.1016/j.polgeo.2010.02.014

Bodon S (2022) 87% of CDC survey participants reported health symptoms from Red Hill water contamination. *Hawai'i Public Radio*. https://www.hawaiipublicradio.org/local-news/2022-05-27/survey-responses-reported-health-symptoms-from-water-navy-red-hill-contamination. Accessed 30 September 2024

Burke MM (2023) 'No clear evidence': Okinawa PFAS contamination comes from U.S. base, Japan's military says. *Stars and Stripes*. https://www.stripes.com/theaters/asia_pacific/2023-05-08/okinawa-pfas-camp-hansen-marine-corps-10056079.html. Accessed 29 September 2024

The United States-Japan Joint Committee (2015) Cooperation concerning environmental matters. ministry of foreign affairs. https://www.mofa.go.jp/mofaj/area/usa/sfa/rem_02.html. Accessed 19 March 2024

Cresswell T (2004) Place: A short introduction. Blackwell Publishing, Hoboken, NJ

Davis S (2015) The empires' edge: militarization, resistance, and transcending hegemony in the Pacific. The University of Georgia Press, Athens, GA

Dowler L (2012) Gender, militarization and sovereignty. Geogr Compass 6(8):490–499. https://doi.org/10.1111/j.1749-8198.2012.00509.x

Elden S (2010) Land, terrain, territory. Prog Hum Geogr 34(6):799–817. https://doi.org/10.1177/0309132510362603

Elden S (2007) Governmentality, calculation, territory. Environ Planning D, Soc & Space 25(3):562–580. https://doi.org/10.1068/d428t

Enloe CH (2014) Bananas, beaches and bases: making feminist sense of international politics, 2nd edn. University of California Press, Berkeley

Environmental Protection Agency (2024) About the red hill fuel releases. Environ Prot Agency. https://www.epa.gov/red-hill/about-fuel-releases#:~:text=In%20January%202014,%20during%20the%20course%20of%20refilling%20Red%20Hill. Accessed 30 September 2024

Flint C (2021) Seapower, geostrategic relations, and islandness: the World War II destroyers for bases deal. Isl Stud J 16(1):271–291. https://doi.org/10.24043/isj.139

Gillem ML (2007) America town: building the outposts of empire. University of Minnesota Press, Chicago

Ginoza A (2012) Space of "Militourism": Intimacies of U.S. and Japanese Empires and Indigenous sovereignty in Okinawa. Int J Okinawan Stud 3(1):7–23

Gonzalez VV (2013) Securing paradise: tourism and militarism in Hawai'i and the Philippines. Duke University Press, Durham, NC

Goulait C (2015) Fuel rebalance. defense logistics agency. https://www.dla.mil/About-DLA/News/News-Article-View/Article/630725/fuel-rebalance/#:~:text=The%20Philippines%E2%80%99%20Subic%20Bay%20facility%20is%20new%20and%20a%20good. Accessed 1 October 2024

Grydehøj A, Bevacqua ML, Chibana M, Nadarajah Y, Simonsen A, Su P, Wright R, Davis S et al (2021) Practicing decolonial political geography: Island perspectives on neocolonialism and the China threat discourse. Polit Geogr 85:1–11. https://doi.org/10.1016/j.polgeo.2020.102330

Haraway D (1988) Situated knowledges: The science question in feminism and the privilege of partial perspective. Fem Stud 14(3):575–599. https://doi.org/10.2307/3178066

Higa H (2022) Toxin blood levels up to 14 times national average among Okinawa residents near U.S. bases. Mainich Shimbun. https://mainichi.jp/english/articles/20221105/p2a/00m/0na/005000c. Accessed 29 September 2024

Ichino K (2023) PFASの一種「PFHxS」を製造使用禁止へ 6月から輸入できず [A PFAS Compound, PFHxS, will be baned for production and import from June]. Asahi Shimbun Digit. https://www.asahi.com/articles/ASRCX3TYPRCXULBH003.html. Accessed 28 September 2024

Jedra C (2021) Amid 'Political Concerns,' Navy Kept Quiet About Red Hill Pipeline Leaking Into Pearl Harbor. Civil Beat. https://www.civilbeat.org/2021/10/amid-political-concerns-navy-kept-quiet-about-red-hill-pipeline-leaking-into-pearl-harbor/#:~:text=But%20an%20active%20leak%20from%20an%20in-use%20pipeline%20would%20reflect. Accessed 30 September 2024

Kajihiro K (2013) Moananuiākea or 'American Lake'? contested histories of the US 'Pacific Pivot'. In: Broudy D, Simpson S, Arakaki M (eds) Under occupation: resistance and struggle in a militarised Asia-Pacific (pp 126–169). Newcastle: Cambridge Scholars Publishing

Kim S (2021) Navy objects to the recommendation it should drain Red Hill fuel tanks. Hawai'i Public Radio. https://www.hawaiipublicradio.org/local-news/2021-12-30/navy-objects-to-the-recommendation-it-should-drain-red-hill-fuel-tanks. Accessed 1 October 2024

Knodell K (2024) Redistribution of Red Hill fuel sparks controversy in Philippines. Starts and Strips. https://www.stripes.com/branches/navy/2024-02-03/hawaii-navy-red-hill-fuel-controversy-philippines-12887832.html. Accessed 1 October 2024

Kuper K (2024) From strategic denial to strategic reclamation: "Mothballed" micronesia. Security Context. https://www.securityincontext.org/posts/from-strategic-denial-to-strategic-reclamation-mothballed-micronesia. Accessed 28 September 2024

Lutz C (2006) Empire is in the details. Am Ethnol 33(4):593–611. https://doi.org/10.1525/ae.2006.33.4.593

Lutz C (2009) The bases of empire: the global struggle against U.S. military posts. New York: New York University Press

McCullough S (2021) Navy says it will comply with state order to remove fuel from Red Hill. Hawai'i Public Radio. https://www.hawaiipublicradio.org/local-news/2022-01-11/navy-will-comply-with-hawaii-order-to-defuel-red-hill-facility. Accessed 1 October 2024

Ministry of Environment (2024). PFOS、PFOA に関するQ&A集 [Q&A regarding PFOS and PFOA]. Ministry of Environment. https://www.env.go.jp/water/pfas/pfas.html. Accessed 28 September 2024

Ministry of Internal Affairs and Communications (2009). 一般戦災死没者の追悼. [Mourning for those who lost their lives in the battle]. Ministry of Internal Affairs and Communications. https://www.soumu.go.jp/main_sosiki/daijinkanbou/sensai/situation/state/okinawa_04.html. Accessed 19 March 2024

Mitchell J (2020) PFAS contamination from U.S. military facilities in Mainland Japan and Okinawa. Asia-Pac J 18(16):1–19

Mitchell J (2023) In Hawai'i, water chief joins PFAS protests against military, urges Okinawans to speak out, too. Okinawa Times. https://www.okinawatimes.co.jp/articles/-/1148885. Accessed 30 September 2024

Mori, K (2022) Resilience of the community against environmental pollution: the knowledge production process of local activism on PFAS contamination on U.S. military bases in Okinawa. Okinawan J Isl Stud 3(2):159–175

Mountz A (2013) Political geography I: Reconfiguring geographies of sovereignty. Prog Hum Geogr 37(6):829–841. https://doi.org/10.1177/0309132513479076

Mountz A (2015) Political geography II: islands and archipelagos. Prog Hum Geogr 39(5):636–646. https://doi.org/10.1177/0309132514560958

Navy Region Hawaii. (n.d). About red hill. America's Navy. https://cnrh.cnic.navy.mil/Operations-and-Management/Red-Hill/About-Red-Hill/. Accessed 29 September 2024

Okinawa Prefectural Government (2018) Brochure (U.S. Military Base). Okinawa prefectural government, Washington D.C. Office. https://dc-office.org/brochure. Accessed 19 March 2024

Okinawa Prefectural Government (2024a) 令和4年度(冬季)有機フッ素化合物環境中残留実態調査の結果について [Results of Environmental Survey of PFAS. 2022 Winter]. Okinawa Prefecture. https://www.pref.okinawa.lg.jp/kurashikankyo/kankyo/1004418/1018739/1021802.html. Accessed 24 February 2024

Okinawa Prefectural Government (2024b) 令和4年度土壌中のPFOS等調査結果 [Results of soil survey regarding PFAS]. Okinawa Prefecture. https://www.pref.okinawa.jp/kurashikankyo/kankyo/1004750/1004804/1004810.html. Accessed 22 February 2024

Okinawa Times (2016) 嘉手納基地が原因か　北谷浄水場から汚染物質　県は米軍に照会 [Polluted materials detected at the Chatan purification plant, potentially the source on the Kadena Air Base. The Prefecture to inquire the U.S. Military]. Okinawa Times. https://www.okinawatimes.co.jp/articles/-/22914. Accessed 28 September 2024

Oshiro A (2023) From "footprint" to relationships: impacts of U.S. military base on Okinawa. Sociol Compass 18(1):1–14. https://doi.org/10.1111/soc4.13099

Osorio JK (2002) Dismembering lāhui: a history of the Hawaiian nation to 1887. University of Hawai'i Press, Hawai'i

Rodenbiker J (2023) Ecological militarization: engineering territory in the South China Sea. Polit Geogr 106:102932. https://doi.org/10.1016/j.polgeo.2023.102932

Ronström O (2012) Finding their place: islands as locus and focus. Cult Geogr 20(2):153–165. https://doi.org/10.1177/1474474012445446

Sack R (1983) Human territoriality: a theory. Ann Assoc Am Geogr 73(1):55–74. https://doi.org/10.1111/j.1467-8306.1983.tb01396.x

Sesoko S (2023) PFAS血中濃度、金武と北谷の受検者の6割が米学術機関の目安値超える　市民団体調査 沖縄 [A Local Group conducted a survey in Okinawa. 60 percent of the samples from Kin and Chatan exceeded the standard level set by a U.S. academic institution]. *Ryukyu Shimpo*. https://ryukyushimpo.jp/news/entry-1664113.html. Accessed 29 September 2024

Steinberg PE (2009) Sovereignty, territory, and the mapping of mobility: a view from the outside. Ann Assoc Am Geogr 99(3):467–495. https://doi.org/10.1080/00045600902931702

Teaiwa T (2011) Bleeding boundaries: gendered analyses of militarism in the western pacific. Asia Pac Viewp 52(1):1–4. https://doi.org/10.1111/j.1467-8373.2011.01441.x

Vine D (2011) Island of shame: the secret history of the U.S. military base on Diego Garcia, 1st edn. Princeton University Press, Princeton

Woodward R (2014) Military landscapes: agendas and approaches for future research. Prog Hum Geogr 38(1):40–61. https://doi.org/10.1177/0309132513493219

MIX
Papier aus verantwortungsvollen Quellen
Paper from responsible sources
FSC® C105338

If you have any concerns about our products,
you can contact us on
ProductSafety@springernature.com

In case Publisher is established outside the EU,
the EU authorized representative is:
Springer Nature Customer Service Center GmbH
Europaplatz 3, 69115 Heidelberg, Germany

Printed by Libri Plureos GmbH
in Hamburg, Germany